SOLUTIONS MANUAL

PHYSICAL CHEMISTRY

Principles and Applications in Biological Sciences

FOURTH EDITION

TINOCO · SAUER · WANG · PUGLISI

Upper Saddle River, NJ 07458

Editor in Chief: John Challice
Project Manager: Kristen Kaiser
Executive Managing Editor: Kathleen Schiaparelli
Assistant Managing Editor: Dinah Thong
Production Editor: Barbara A. Till
Supplement Cover Manager: Paul Gourhan
Supplement Cover Designer: PM Workshop Inc.
Manufacturing Buyer: Lisa McDowell

Prentice Hall

Upper Saddle River, NJ 07458

Printed in the United States of America

10 9 8 7 6 5 4 3

ISBN 0-13-026607-8

Prentice-Hall International (UK) Limited, London
Prentice-Hall of Australia Pty. Limited, Sydney
Prentice-Hall Canada, Inc., Toronto
Prentice-Hall Hispanoamericana, S.A., Mexico City
Prentice-Hall of India Private Limited, New Delhi
Pearson Education Asia Pte. Ltd., Singapore
Prentice-Hall of Japan, Inc., Tokyo
Editora Prentice-Hall do Brazil, Ltda., Rio de Janeiro

Contents

CHAPTER 2

1. a) $q_p = \Delta H = (1\ \text{L})(0.9928\ \text{kg/L})(2447\ \text{kJ/kg})$

$q_p = \underline{2443\ \text{kJ}}$

b) $\Delta T = \dfrac{(2443\ \text{kJ})}{(60\ \text{kg})[4.18\ \text{kJ/(K kg)}]}$

$\Delta T = \underline{9.7\ \text{K}}$

c) $C_{12}H_{22}O_{11}(s) + 12O_2(g) = 12CO_2(g) + 11H_2O(l)$

$\Delta H = 11(-285.83) + 12(-393.51) - (-2222.1)$

$\Delta H = -5644\ \text{kJ/mol}$

$\text{mol. wt. of sucrose} = 342.31\ \text{g/mol}$

$\text{wt.} = \dfrac{(2443\ \text{kJ})(342.31\ \text{g/mol})}{5644\ \text{kJ/mol}}$

$\text{wt.} = \underline{148.2\ \text{g}}$

2. a) $\text{Weight of air needed} = (1\text{kg})(44/12)/(4.6\times10^{-4}) = 7.97\times10^{3}\ \text{kg}$

$\text{"Molecular weight" of air} = (0.8)(0.028) + (0.2)(0.032) = 0.0288\text{kg/mol}$

$\text{"Moles" of air needed} = (7.97\times10^{3})/(0.0288) = 2.77\times10^{5}$

$\text{Volume of air needed} = (2.77\times10^{5})(0.08205)(298)/(1) = \underline{6.8\times10^{6}\ \text{L}}$

b) $1\ \text{atm} = 1.033\times10^{4}\ \text{kg m}^{-2}$

$\text{Wt of } CO_2 \text{ over } 1\ \text{m}^2 = (1.03\times10^{4}\ \text{kg})(4.6\times10^{-4}) = 4.75\ \text{kg } CO_2$

$\text{Wt of C} = (4.75\ \text{kg})(12.0)/(44.0) = \underline{1.30\ \text{kgC}}$

c) $(1.30\ \text{kg C})/(1.00\ \text{kg C y}^{-1}) = \underline{1.30\ \text{y}}$

3. a) $w = mgh = (10\ \text{kg})(9.81\ \text{m s}^{-2})(10\ \text{m}) = 981\ \text{kg m}^2\ \text{s}^{-2} = \underline{981\ \text{J}}$

b) $w = EIt = (6.0\ \text{V})(5.5\ \text{A})(2\ \text{h})(3600\ \text{s/h})\left(\dfrac{\text{J/s}}{\text{VA}}\right) = \underline{2.38\times10^{5}\ \text{J}}$

c) $k = \dfrac{\text{force}}{(x - x_0)} = \dfrac{5.00\ \text{N}}{(0.105\ \text{m} - 0.100\ \text{m})} = 1.00\times10^{3}\ \text{N/m}$

$$w = \frac{k}{2}(x - x_0)^2 = \frac{(1.00 \times 10^3 \text{ N/m})(0.01 \text{ m})^2}{2} (\text{J/N m})$$

$$w = \underline{0.050 \text{ J}}$$

$$\frac{\Delta d}{d_{\text{surface}}} \times 100\% = \underline{1.2\%}$$

d) $$w = -P(V_2 - V_1) = \frac{(-1.00 \text{ atm})(2 \text{ L})(8.314 \text{ J})}{0.08205 \text{ L atm}}$$

$$w = -203 \text{ J}$$

e) $$w = -203 \times 10^{-6} \text{ J}$$

f) $$w = -nRT \ln\frac{V_2}{V_1} = P_1V_1 \ln\frac{V_2}{V_1}$$

$$w = [-(1.00 \text{ atm})(1.00 \text{ L}) \ln 3](\frac{8.314 \text{ J}}{0.08205 \text{ L atm}})$$

$$w = -111 \text{ J}$$

4. a)

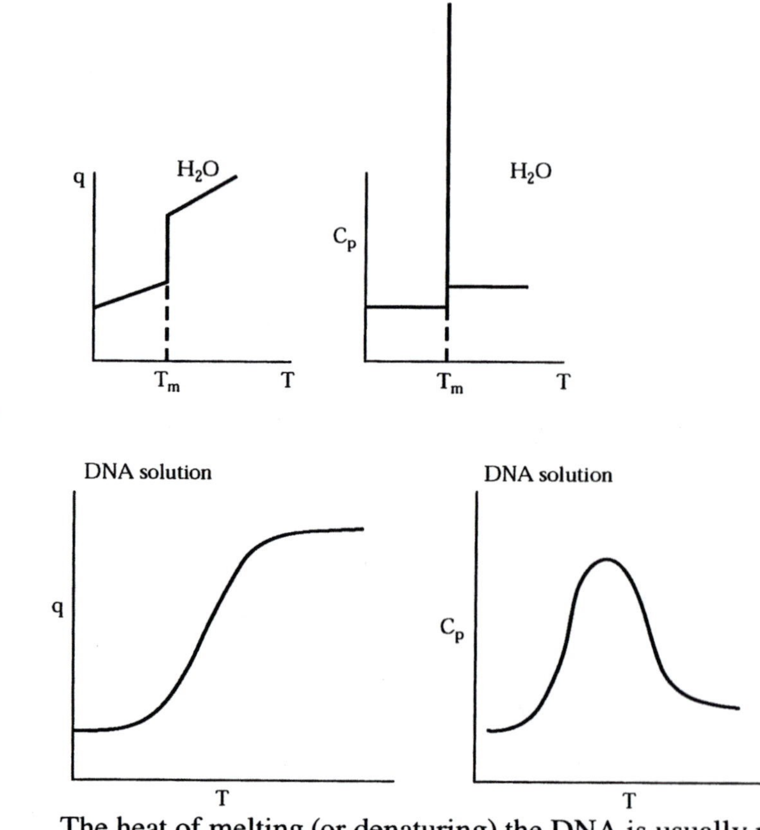

The heat of melting (or denaturing) the DNA is usually measured by integrating the heat capacity curve between an initial and final temperature assigned to the transition. This is equal to the amount of heat necessary to raise the temperature of the solution from the initial to the final temperature. The heats of the DNA solution are measured relative to the pure buffer solution.

5. a) Wt of H_2O over 1 m^2 of ocean floor $= (2500 \text{ m})(10^3 \text{ L m}^{-3})(1.025 \text{kg L}^{-1})$

$$= 2.5\times10^6 \text{ kg m}^{-2}$$

$$P = (2.5\times10^6 \text{ kg m}^{-2})(1 \text{ atm}/1.033\times10^4 \text{ kg m}^{-2}) = \underline{248 \text{ atm}}$$

b) $$\frac{\Delta d}{d_o} = \frac{-\Delta V}{V_o} = \beta\Delta P$$

$$\frac{\Delta d}{d_o} = (49.5\times10^{-6} \text{atm}^{-1})(247 \text{atm}) = 0.0122 = 1.22\%$$

c) $$V_{\text{bottom}} = (10 \text{ L})(277/293)(1/248) = 0.0381 \text{ L}$$

$$\frac{\Delta d}{d_{\text{surface}}} = \frac{V_{\text{surface}} - V_{\text{bottom}}}{V_{\text{bottom}}} = \frac{10 - 0.0381}{0.0381}$$

$$\frac{\Delta d}{d_{\text{surface}}} \times 100\% = \underline{2.61\times10^4\%}$$

d) $$n = PV/RT = (1)(10)/(0.08205)(293) = 0.416 \text{ mol}$$

$$(P + n^2 a/V^2)(V - nb) = nRT$$

$$[248 + (0.416)^2(1.36)V^{-2}][V - (0.416)(3.183\times10^{-2})] = (0.416)(0.08205)(277)$$

$$248V - 3.20 + 0.235V^{-1} - 3.12\times10^{-3}V^{-2} = 9.45$$

$$248V^3 - 12.65V^2 + 0.235V - 3.12\times10^{-3} = 0$$

$$V = 0.0344 \text{ L}$$

$$\frac{\Delta d}{d_{\text{surface}}} \times 100\% = \left(\frac{10 - 0.0344}{0.0344}\right) 100 = 2.90\times10^4\%$$

6. Moles of $H_2O = (100 \text{ g})/(18.0 \text{ g mol}^{-1}) = 5.56 \text{ mol}$

a) $q = (5.56 \text{ mol})(75.4 \text{ J K}^{-1} \text{ mol}^{-1})(100 \text{ K}) = \underline{41.9 \text{ kJ}}$

b) $q = \Delta H_{\text{fus}} = (100 \text{ g})(-333 \text{ J g}^{-1}) = \underline{-33.3 \text{ kJ}}$

c) $q = \Delta H_{\text{vap}} = (100 \text{ g})(2257 \text{ J g}^{-1}) = \underline{226 \text{ kJ}}$

7. $V_1 = (1\text{ mol})\left(0.08205\text{ L atm K}^{-1}\text{ mol}^{-1}\right)(300\text{ K})/(1\text{ atm}) = 24.6\text{ L}$

$V_2 = V_1(600\text{ K})/(300\text{ K}) = 49.2\text{ L}$

a) $w_p = -P(V_2 - V_1) = -(1\text{ atm})(49.2 - 24.6)\text{ L}\left(101.3\text{ J L}^{-1}\text{ atm}^{-1}\right)$

$= \underline{-2.49\text{ kJ}}$

b) $\Delta E = C_v \Delta T = \left(20.8\text{ J K}^{-1}\text{ mol}^{-1}\right)(300\text{ K}) = \underline{6.24\text{ kJ}}$

$\Delta H = \Delta E + \Delta(PV) = 6.24 + 2.49 = \underline{8.73\text{ kJ}}$

c) $q_p = \Delta H = \underline{8.73\text{ kJ}}$

8. a) For an isothermal expansion $V_2/V_1 = P_2/P_1$

$T_2 = \underline{300\text{ K}}$

$\Delta E = \underline{0}$; $\Delta H = \underline{0}$

$q_T = -w_T = nRT\ln(V_2/V_1) = (1\text{ mol})(8.314\text{ J K}^{-1}\text{ mol}^{-1})(300\text{ K})\ln 10$

$= \underline{5.74\text{ kJ}}$

b) $q = 0$

$w = -P_{ex}(V_2 - V_1) = \Delta E = n\overline{C}_v(T_2 - T_1)$

$$(1\text{ atm})\left(\frac{nRT_1}{P_1} - \frac{nRT_2}{P_2}\right) = n\overline{C}_v(T_2 - T_1)$$

$\overline{C}_v T_2 + RT_2 = \overline{C}_v T_1 + RT_1/10 = \left(\overline{C}_v + R/10\right)T_1$

$$T_2 = \frac{\left(\overline{C}_v + R/10\right)}{\overline{C}_v + R}T_1 = \frac{1.6R}{2.5R}(300\text{ K}) = \underline{192\text{ K}}$$

$\Delta E = w = (1\text{ mol})(3R/2)(192 - 300) = \underline{-1.347\text{ kJ}}$

$\Delta H = n\overline{C}_p(T_2 - T_1) = (5R/2)(192 - 300) = -2.245\text{ kJ}$

c) $w = 0$; $q = 0$; $\Delta E = 0$; $T_2 = 300\ K$; $\Delta H = 0$

9. a) $V_1 = (1\text{ mol})\left(0.08205\text{ L atm K}^{-1}\text{ mol}^{-1}\right)(298\text{ K})/(10\text{ atm}) = 2.445\text{ L}$

$V_2 = 10V_1 = 24.45\text{ L}$

$w = -P_{ex}(V_2 - V_1) = -(1\text{ atm})(24.45 - 2.45)\text{ L}\left(101.3\text{ J L}^{-1}\text{ atm}^{-1}\right) = -2.23\text{ kJ}$

$\Delta E = \Delta H = \underline{0}$

$q_T = -w_T = \underline{2.23\text{ kJ}}$

b) $w = 0$

$q = \Delta E = n\overline{C}_V \Delta T = (1 \text{ mol})(5R/2)(373 - 298) = \underline{1.559 \text{ kJ}}$

$\Delta H = n\overline{C}_P \Delta T = (1 \text{ mol})(7R/2)(75) = \underline{2.182 \text{ kJ}}$

c) $P_2 = P_1(T_2/T_1) = (1 \text{ atm})(373\text{K})/(298\text{K}) = \underline{1.252 \text{ atm}}$

d) Final temperature will be lower. By expanding, the gas will do work, $w < 0$. Therefore $\Delta E = w < 0$ and $T_2 < T_1$.

10. a) $q = 0$, adiabatic

$w < 0$, expansion

$\Delta E = w < 0$

$\Delta H < 0$

b) $\Delta E = \Delta H = 0$ since $\Delta T = 0$

$w < 0$, expansion

$q = -w > 0$

c) $q = 0$, adiabatic

$w = 0$, expansion against zero pressure

$\Delta E = q + w = 0$

$\Delta H = \Delta E + R\Delta T = 0$

d) $q > 0$, heat of vaporization

$w < 0$, expansion against pressure

$\Delta H = q_P > 0$

$\Delta E = \Delta H - \Delta(PV) > 0; \Delta H >> \Delta(PV) \cong n_g RT$

e) $q < 0$, exothermic reaction

$w = 0$, closed bomb, $\Delta V = 0$

$\Delta E = q_V < 0$

$\Delta H = q_V + \Delta(PV) < 0; \Delta(PV) = -(n_{H_2} + n_{O_2})RT < 0$

11. a) $V_2 = (1)(0.08205)(373)/(1) = 30.6\text{L}$

V_1 is negligible

$w_P = -P\Delta V = -(1\text{atm})(30.6\text{ L})(101.3\text{ J L}^{-1}\text{ atm}^{-1})$

$= \underline{-3.10\text{ kJ}}$

$q = 40.66\text{ kJ mol}^{-1}$, from Table 2.2

$\Delta E = q + w = \underline{37.56\text{ kJ mol}^{-1}}$

$\Delta H = q_P = \underline{40.66\text{ kJ mol}^{-1}}$

b) $w = w_1 + w_2 = 0$, since $P_{ex} = 0$ in step 1 and $\Delta V = 0$ in step 2

$\Delta E = \underline{37.56\text{ kJ mol}^{-1}}$, path independent

$q = \Delta E = \underline{37.56\text{ kJ mol}^{-1}}$

$\Delta H = \underline{40.66\text{ kJ mol}^{-1}}$, path independent

12. a) $q = C(T_2 - T_1) = (18\text{ g})(2.113\text{ J K}^{-1}\text{ g}^{-1})(20\text{ K}) = \underline{761\text{ J}}$

b) $q = (18\text{ g})(333.4\text{ J g}^{-1}) = \underline{6001\text{ J}}$

c) Per degree temperature drop, 180 g H_2O (l) loses

$(180\text{ g})(4.18\text{ J g}^{-1}) = 752.4\text{ J}$

$761\text{ J} + 6001\text{ J} + (18\text{ g})(4.18\text{ J K}^{-1}\text{g}^{-1})(T_f - 0) = (180\text{ g})(4.18\text{ J K}^{-1}\text{g}^{-1})(20 - T_f)$

$$T_f = \frac{(180)(4.18)(20) - 761 - 6001}{(198)(4.18)} = \underline{10.0°\text{C}}$$

13. a) $\Delta T = q_P / C_P = -400\text{ J}/(1.874\text{ J K}^{-1}\text{g}^{-1})(18.0\text{ g})$

$= -11.9°\text{C}$

$T_f = 125 - 11.9 = \underline{113.1°\text{C}}$

There is no phase change, so

$$\Delta V = nR\Delta T / P = \frac{1(0.08205)(-11.9)}{1}$$

$\Delta V = \underline{-0.98L}$

b) $\Delta T = -(400/18)/4.18 = -5.32°\text{C}$

$T_f = 100 - 5.32 = \underline{94.68°\text{C}}$

$V_f(l) = (18\text{ g})/(0.96\text{ g ml}^{-1}) = 18.75\text{ ml}$. No phase change; ΔV is negligible for the liquid water.

c) $\Delta n_g = (-400\text{ J})/(40{,}660\text{ J mol}^{-1}) = \underline{-9.84\times 10^{-3}\text{ mol}}$

Δn_g mol water vapor will condense to liquid

No temperature change

$\Delta V = \Delta nRT/P = (-9.84\times 10^{-3})((0.08205)(373)/1$

$\Delta V = \underline{-0.30 L}$

d) Part (a), because it has the greatest $\Delta V; w > 0$

14. a) $\Delta H^{\circ}_{298} = \Delta H_1 + \Delta H_2 + \Delta H_3$

$= (18.0\text{ g})(4.18\text{ J K}^{-1}\text{ g}^{-1})(75\text{ K}) + 40{,}660\text{ J} + (18.0\text{ g})\times(1.874\text{ J K}^{-1}\text{g}^{-1})(-75\text{ K})$

$= \underline{43{,}773\text{ J}}$; assumes heat capacities are constant over the temperature range

b) $\Delta H^{\circ}_{298} = \Delta H_1 + \Delta H_2 + \Delta H_3$

$= 0 + (18.0\text{ g})(2436\text{ J g}^{-1}) + 0$

$= \underline{43{,}848\text{ J}}$; assumes ΔH changes linearly with T between 20ÞC and 40ÞC, and that ΔH is not pressure dependent

Values should be the same as ΔH is independent of path.

c) $\Delta H^{\circ}_{298} = -241.82 - (-285.83) = 44.01\text{ kJ} = \underline{44{,}010\text{ J}}$

d) $\%\text{deviation} = (44{,}010 - 43{,}773)(100)/44{,}010 = \underline{0.54\%}$

The result of part (c) is the most accurate, because no approximations are made.

15. a) $q = 0$, thermally insulated

$w = 0$, assuming no volume change because decrease in volume of hot copper equals increase in cool copper.

$\Delta E = 0$, energy is conserved

$\Delta H = 0$

b) $q = 0$, thermally insulated

$w > 0$, mechanical work done on the system

$\Delta E = w > 0$

$\Delta H > 0, \Delta(PV) \sim 0,$ negligible expansion of the system

c) $\left.\begin{array}{l} q=0 \\ w=0 \\ \Delta E=0 \\ \Delta H=0 \end{array}\right\}$ assumes that the gases are ideal in their behavior

16. a) First Law of Thermodynamics – no restrictions

b) Constant pressure process

c) Constant pressure; enthalpy is a linear function of temperature. i.e. $C =$ constant; $\Delta H/\Delta T = dH/dT$

d) Ideal gases, where $\Delta(PV) = \Delta(nRT)$

e) Gases only; limited range of P, T

f) Constant P_{ext}, expansion

17. $(4 \text{ kJ h}^{-1})/(1 \text{ kg})(4.18 \text{ kJ K}^{-1} \text{ kg}^{-1}) \cong 1 \text{ degree h}^{-1}$

A temperature rise of 5 ˚F or 3 ˚C would correspond to a state of high fever. Thus, three hours would be an upper limit.

18. $n = PV/RT = (1 \text{ atm})(0.5 \text{ L})/(0.082 \text{ L atm K}^{-1} \text{ mol}^{-1})(293 \text{ K}) = 0.021 \text{ mol}$

$\Delta H = n\overline{C}_P(T_2 - T_1) = (0.021)(30 \text{ J K}^{-1})(17 \text{ K}) = 10.6 \text{ J}$

Heat loss from breathing $= (10.6 \text{ J})(12)(60)(24) = 180 \text{ kJ d}^{-1}$ representing about 1.5% of food energy

Metabolic heat $= (80 \text{ kg})(4 \text{ kJ h}^{-1} \text{ kg}^{-1})(24 \text{ h}) = 7.68 \text{ kJ d}^{-1}$ or 0.06% of food energy. At outside temperatures of $-40°\text{C}$, 4.5x more energy will be lost through breathing; 7% of food energy begins to be significant.

19. a) $w = P\Delta V = (1 \text{ atm})(0.5 \text{ L})(101.3 \text{ J L}^{-1} \text{ atm}^{-1}) \cong 50 \text{ J/breath}$

During 24 h, $w = (50 \text{ J})(15{,}000) = \underline{750 \text{ kJ}}$

b) $w = \text{mass g }(\Delta H) = 750 \text{ kJ} = \text{mass }(9.8 \text{ m s}^{-2})(100 \text{m})$

$$\text{mass} = \frac{750{,}000 \text{ kg m}^2 \text{ s}^{-2}}{(9.8 \text{ ms}^{-2})(100 \text{ m})} = \underline{765 \text{ kg}} \cong 0.85 \text{ tons}$$

20. $q_P = \Delta H = 0$

$$\mathrm{wt}(1)\overline{C}_P(1)(T_f - 55) + \mathrm{wt}(s)C_P(s)(10) + \mathrm{wt}(s)\Delta\overline{H}(\text{melting}) + \mathrm{wt}(s)C_p(1)(T_f) = 0$$

$$(0.100\ \text{kg})(4.18\ \text{kJ K}^{-1}\text{kg}^{-1})(T_f\text{K} - 55\ \text{K}) + (0.010\ \text{kg})(2.11\ \text{kJ K}^{-1}\ \text{kg}^{-1})(10\ \text{K})$$

$$+(0.010\ \text{kg})(333.4\ \text{kJ kg}^{-1}) + (0.010\ \text{kg})(4.18\ \text{kJ k}^{-1}\ \text{kg}^{-1})(T_f\ \text{K}) = 0$$

$$0.418T_f - 22.99 + 0.21 + 3.33 + 0.042T_f = 0$$

$$T_f = \frac{19.45}{0.460} = \underline{42.3°\text{C}}$$

21. $$\Delta H^\circ_{298} = 6\Delta\overline{H}^\circ_f(CO_2, g) + 6\Delta\overline{H}^\circ_f(H_2O, l) - \Delta\overline{H}^\circ_f(C_6H_{12}O_6, s) - 6\Delta\overline{H}^\circ_f(O_2, g)$$

$$= 6(-393.509) + 6(-285.830) - (-1274.4) - 0 = \underline{-2801.6\ \text{kJ}}$$

22. a) $$q_p = \Delta H = 2\Delta\overline{H}^\circ_f(\text{ethanol}, l) + 2\Delta\overline{H}^\circ_f(CO_2, g) - \Delta\overline{H}^\circ_f(\text{glucose}, s)$$

$$= 2(-276.98) + 2(-393.509) - (-1274.4) = \underline{-66.6\ \text{kJ}}$$

b) This represents 2.4% of the heat available from the complete combustion.

23. $$\Delta H^\circ = \Delta\overline{H}^\circ_f(H_2O, l) + (1/2)\Delta\overline{H}^\circ_f(O_2, g) - \Delta\overline{H}^\circ_f(H_2O_2, aq)$$

$$\Delta H^\circ = (-285.830) + 0 - (-191.17) = -94.66\text{kJ mol}^{-1}$$

A temperature rise of 0.02 °C requires (0.02 °C)(4.18 kJ kg^{-1})(1 kg L^{-1}) = 0.0836 kJ L^{-1}

$$\text{Minimum detectable concentration} = \frac{0.0836\ \text{kJ L}^{-1}}{94.66\text{kJ mol}^{-1}} = 8.83 \times 10^{-4}\ \text{M}$$

24. a) $$\Delta H^\circ_{298} = (-74.81) + 0 - (-238.57) = \underline{163.76\ \text{kJ}}$$

b) $$\Delta E^\circ = \Delta H^\circ - \Delta n_{\text{gases}}RT = 163.76 - (3/2)(8.314/1000)(298)$$

$$= \underline{160.04\ \text{kJ}}$$

c) $$\Delta H^\circ_{773} = \Delta H^\circ_{298} + \int_{298}^{773} \Delta C_p dT$$

where $\Delta C_P = \overline{C}_P(CH_4) + 1/2\overline{C}_P(O_2) - \overline{C}_P(CH_3OH)$

25. $$\Delta H^o_{298} = \Delta H^o_f(urea, s) + 3\Delta H^o_f(CO_2,g) + 3\Delta H^o_f(H_2O,l) - 2\Delta H^o_f(glycine, s)$$

$$\Delta H^\circ_{298} = (-333.17) + 3(-393.509) + 3(-285.83) - 2(-537.2)$$

$$= \underline{-1296.8 \text{ kJ mol}^{-1}}$$

26. a) $\Delta H^\circ_{298} = 2(-276.98) + 2(-393.509) - (-1274.4) = \underline{-66.6 \text{ kJ mol}^{-1}}$,

(see problem 22a)

b) $$\Delta H^\circ_{298} = 2(-484.1) + 2(-393.509) + 2(-285.830) - (-1274.4)$$

$$= \underline{-1052.5 \text{ kJ mol}^{-1}}$$

c) $\Delta H^\circ_{298} = \underline{-2801.6 \text{ kJ mol}^{-1}}$ (see Prob. 21)

27. $$\Delta H^\circ_{353} = \Delta H^\circ_{298} + \Delta C_p(80 - 25)$$

a) $$\Delta C_p = 2\overline{C}_p(\text{C}_2\text{H}_5\text{OH}) + 2\overline{C}_p(\text{CO}_2) - \overline{C}_p(\text{C}_6\text{H}_{12}\text{O}_6)$$

$$= 2(111.5) + 2(37.1) - (225) = 72.1 \text{ J K}^{-1}\text{mol}^{-1}$$

$$\Delta H^\circ_{353} = -66.6 + (72.1/1000)(55) = \underline{-62.6 \text{ kJ mol}^{-1}}$$

b) $$\Delta C_p = 2\overline{C}_p(\text{CH}_3\text{COOH}) + 2\overline{C}_p(\text{CO}_2) + 2\overline{C}_p(\text{H}_2\text{O}) - 2\overline{C}_p(\text{O}_2) - \overline{C}_p(\text{C}_6\text{H}_{12}\text{O}_6)$$

$$= 2(123.5) + 2(37.1) + 2(75.4) - 2(29.4) - (225)$$

$$= 188.2 \text{ J K}^{-1}\text{ mol}^{-1}$$

$$\Delta H^\circ_{353} = -1052.5 + (188.2/1000)(55) = \underline{-1042.1 \text{ kJ mol}^{-1}}$$

28. a) $$\Delta H^\circ_{298} = \Delta\overline{H}^\circ_f(\text{sucrose,s}) + 12\Delta\overline{H}^\circ_f(\text{O}_2,g) - 12\Delta\overline{H}^\circ_f(\text{CO}_2,g) - 11\Delta H^\circ_f(H_2\text{O,l})$$

$$= (-2222.1) + 0 - 12(-393.509) - 11(-285.830)$$

$$= \underline{5644.1 \text{ kJ mol}^{-1}}$$

b) $$20 \text{ kg sucrose}/(\text{hectare h}) = (5644.1 \text{ kJ mol}^{-1})(0.3423 \text{ kg mol}^{-1})^{-1}(20 \text{ kg})$$

$$= 329{,}800 \text{ kJ h}^{-1} = \underline{91.6 \text{ kW (hectare)}^{-1}}$$

c) $$\text{Stored energy} = [(91.6 \text{ kW hectare}^{-1})(10^{-4} \text{ hectare m}^{-2})/1 \text{ kW m}^{-2}]100$$

$$= \underline{0.92\%}$$

29. a) $H_2(g) + 1/2O_2(g) + H_2O(l)$

$$\Delta H^\circ_{298} = \Delta H^\circ_f(H_2O,l) = (-285.830 \text{ kJ mol}^{-1})/(2.016 \text{ g } H_2 \text{ mol}^{-1})$$

$$= \underline{-141.8 \text{ kJ}(gH_2)^{-1}}$$

b) $C_8H_{18}(g) + 25/2O_2 \rightarrow 8CO_2(g) + 9H_2O(l)$

$$\Delta H^\circ_{298} = [-(-208.45) - 0 + 8(-393.509) + 9(-285.830)]/(114 \text{ g}C_8H_{18}/\text{mol})$$

$$= \underline{-48.25 \text{ kJ}(gC_8H_{18})^{-1}}$$

c) $\Delta H^\circ(H_2)/\Delta H^\circ(C_8H_{18}) = 141.8/48.25 = \underline{2.9}$

Thus H_2 is nearly 3 times more energetic as a fuel than octane on a weight basis.

30. a) $\Delta H^\circ = \Delta H^\circ_f(\text{Ala}) + \Delta H^\circ_f(CO_2) - \Delta H^\circ_f(\text{Asp})$

$$= (-562.7) + (-393.509) - (-973.37)$$

$$= \underline{17.2 \text{ kJ mol}^{-1}}\text{, absorbed}$$

b)

$$\begin{array}{ccc} \text{Asp (50°C)} & \dashrightarrow & \text{Ala (50°C)} + CO_2 \text{ (50°C)} \\ \searrow & & \nearrow \quad \nearrow \\ \text{Asp (25°C)} & \longrightarrow & \text{Ala (25°C)} + CO_2 \text{ (25°C)} \end{array}$$

Heat capacities needed

$$\Delta H^\circ_{323} = \Delta H^\circ_{298} + \Delta C_p (50 - 25)$$

where $\Delta C_p = \overline{C}_p(\text{Ala}) + \overline{C}_p(CO_2) - \overline{C}_p(\text{Asp})$

31. To measure ΔH°_{298} the reactants could be put into a calorimeter at 25 °C and a small amount of an enzyme catalyst added. The heat evolved per mole at constant pressure is ΔH°_{298}. To estimate ΔE°_{298} the volume change of the reaction must be determined. The change in volume can be measured in a dilatometer, or partial molal volumes can be used to calculate the change in volume. For practical purposes the volume change is negligible, since no gases are involved, and $\Delta E^\circ_{298} = \Delta H^\circ_{298}$.

32. a) $(22 \text{ kWh day}^{-1})(3600 \text{ kJ kWh}^{-1}) = \underline{7.92 \times 10^4 \text{ kJ day}^{-1}}$

b) $(1 \text{ kW m}^{-2})(0.1)(5 \text{ h}) = 0.5 \text{ kWh m}^{-2}$

$$A = (22 \text{ kWh})/(0.5 \text{ kWh m}^{-2}) = \underline{44 \text{ m}^2}$$

33. a) $$\Delta H^\circ_{298} = 2\Delta \overline{H}^\circ_f (H_2O, g) + \Delta \overline{H}_f (O_2, g) - \Delta \overline{H}^\circ_f (H_2O_2, g)$$
$$= 2(-241.818) + 0 - 2(-133.18) = \underline{-217.28 \text{ kJ mol}^{-1}}$$

b)

$$H_2O_2 \dashrightarrow 2OH$$

$H_2O_2 \rightarrow H_2O + 1/2\ O_2$: $-1/2\ (217.28)$

$H_2O + 1/2\ O_2 \rightarrow HO + H + O$: $+\ D(OH) + 1/2\ D\ (O_2)$

$HO + H + O \rightarrow 2OH$: $-D(O\text{–}H)$

$$D(O-O) = -1/2(217.28) + 1/2(498.3) = \underline{140.5 \text{ kJ mol}^{-1}}$$

c) $$\Delta H^\circ_{298} = 2(-285.830) + 0 - 2(-191.17) = \underline{-189.32 \text{ kJ mol}^{-1}}$$

d) $$\text{Heat evolved} = 1/2(189.32 \text{ kJ/mol } H_2O_2)(0.01 \text{ mol kg}^{-1}) = 0.947 \text{ kJ kg}^{-1}$$

$$\text{Temperature rise} = (0.947 \text{ kJ kg}^{-1})/(4.18 \text{ kJ K}^{-1} \text{ kg}^{-1}) = 0.226 \text{ degrees}$$

Final temperature <u>25.226°C</u>

34. a) $$8\ C(\text{graphite}) + 9H_2(g) \rightarrow C_8H_{18}(g)$$

$$\Delta H^\circ_f = 8(716.7) + 9(436.0) - 7(344) - 18(415) = \underline{-220 \text{ kJ mol}^{-1}}$$

Published value = –208.45 kJ mol^{-1}

Chief discrepancy lies in using average bond dissociation energies for D(C–C) and D(C–H)

b) 10 C (graphite) + 4 H_2 (g) →

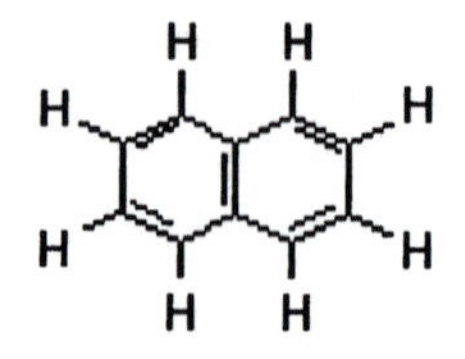

$$\Delta H^\circ_f = 10\Delta H_f(C) + 4D(H-H) - 6D(C-C) - 5D(C=C) - 8D(C-H)$$
$$= 10(716.7) + 4(436.0) - 6(344) - 5(615) - 8(415)$$
$$= \underline{452 \text{ kJ mol}^{-1}}$$

Published value 150.96 kJ mol^{-1}

Chief discrepancy arises from neglect of resonance energy.

c) $C(\text{graphite}) + H_2(g) + 1/2O_2(g) \rightarrow H_2C = O(g)$

$$\Delta H_f^\circ = \Delta H_f(C) + D(H-H) + 1/2D(O_2) - 2D(C-H) - D(C=O)$$

$$= 716.7 + 436.0 + 249.2 - 2(415) - 725 = \underline{-153 \text{ kJ mol}^{-1}}$$

Published value $-115.90 \text{ kJ mol}^{-1}$

Chief discrepancy arises from use of average bond dissociation energies.

d) C (graphite) + H_2 (g) + O_2 (g) $\rightarrow$ formic acid

$$\Delta H_f^\circ = \Delta H_f(C) + D(H-H) + D(O_2) - D(C-H) - D(C=O) - D(C-O) - D(O-H)$$

$$= 716.7 + 436.0 + 498.3 - 415 - 725 - 350 - 463$$

$$= \underline{-302 \text{ kJ mol}^{-1}}$$

Published value $-424.76 + 46.15 = -378.61 \text{ kJ mol}^{-1}$

Chief discrepancy comes from neglect of resonance energy of formic acid H—C(O)(O)H

35. The steric repulsion can be estimated by the enthalpy change for the process

cis - 2 - butene $\rightarrow$ trans - 2 - butene

$$\Delta H_{298}^\circ = -(7.0) + (-11.1) = -4.1 \text{ kJ mol}^{-1}$$

by which we see that the trans form has the lower enthalpy by 4 kJ mol^{-1}.

CHAPTER 3

1. a) Path I: Expansion against 1 atm at constant T (400 K), followed by decrease in T to 300 K at constant P

$$w = w_A + w_B = -P_{ex}(V_2 - V_1) - P_{ex}(V_3 - V_2) \text{ where } P_{ex} = 1 \text{ atm}$$
$$= -(1\text{ atm})(V_3 - V_1) = -(1\text{ atm})R[(T_2/P_2) - (T_1/P_1)]$$
$$= -R(300 - 400/2) = -100\,R = \underline{-1247.1\text{J}}$$

$$q = q_A + q_B = -w_A + \overline{C}_P(T_2 - T_1)$$
$$= (1\text{ atm})RT_1(1/P_2 - 1/P_1) + \overline{C}_P(T_2 - T_1)$$
$$= R(400)(1 - 1/2) + (5/2)R(-100) = -50\,R = \underline{-415.7\text{J}}$$

Path II: Decrease in T to 300 K at constant P, followed by expansion against 1 atm at constant T (300 K)

$$w = w_A + w_B = -(2\text{atm})(V_2 - V_1) - 1\text{atm}(V_3 - V_2)$$
$$= -(1\text{atm}(V_3 + V_2 - 2V_1) = -(1\text{atm})R[T_2/P_2 + T_2/P_1 - 2T_1/P_1]$$
$$= -R[300 + 300/2 - 2(400)/2] = -50\,R = \underline{-415.7\text{J}}$$

$$q = q_A + q_B = \overline{C}_P(T_2 - T_1) - w_B$$
$$= \overline{C}_P(T_2 - T_1) + (1\text{atm})R(T_2/P_2 - T_2/P_1)$$
$$= (5/2)R(-100) + R(300)(1 - 1/2) = -100R = \underline{-831.4\text{J}}$$

b) $\Delta E_{\text{I}} = w + q = -100\,R - 50\,R = -150\,R = \underline{-1247.1\text{ J}}$ (from first law)

$$\Delta E_{\text{II}} = w + q = -50\,R - 100\,R = -150\,R = \underline{-1247.1\text{ J}}$$

$$\Delta S_{\text{I}} = \Delta S_A + \Delta S_B = -R\ln(P_2/P_1) + \overline{C}_P \ln(T_2/T_1)$$
$$= R[(5/2)\ln(300/400) - \ln(1/2)]$$
$$= 0.0261R = \underline{-0.217\text{J K}^{-1}}$$

$$\Delta S_{\text{II}} = \Delta S_A + \Delta S_B = \overline{C}_P \ln(T_2/T_1) - R\ln(P_2/P_1)$$
$$= R[(5/2)\ln(300/400) - \ln(1/2)]$$
$$= -0.0261R = \underline{-0.217\text{J K}^{-1}}$$

2. a) Eff $= -w/q_1 = 0.75; q_1 = (100\text{kJ})/(0.75) = \underline{133.3\text{kJ}}$ (Eq. 3.3)

$$q_1 + q_3 = -w; q_3 = -w - q_1 = (100\text{kJ}) - (133.3\text{kJ}) = \underline{-33.3\text{kJ}}$$

q_1 (positive) is heat absorbed by the system at the hot temperature. q_3 (negative) is heat discharged by the system at the low temperature. w (negative) is the work done by the system.

b) $w = +100\,\text{kJ} = -q_1 - q_3 = -q_1 + 0.25q_1$

Thus $q_1 = \underline{-133.3\,\text{kJ}}$ and $q_3 = \underline{+33.3\,\text{kJ}}$

Heat q_3 (positive) is absorbed by the system from a cold reservoir and heat q_1 (negative) is discharged to the hot reservoir when work w (positive) is done on the system. The engine is acting as a refrigerator.

c) $q_1' = -w'/\text{Eff} = (100\,\text{kJ})/(0.80) = \underline{125\,\text{kJ}}$

$q_3' = -w' - q_1' = 100 - 125 = \underline{-25\,\text{kJ}}$

d) Net heat absorbed at $T_{\text{cold}} = q_3 + q_3' = 33.3 - 25 = \underline{8.3\,\text{kJ}}$

Net heat discharged at $T_{\text{hot}} = q_1 + q_1' = -133.3 + 125 = \underline{-8.3\,\text{kJ}}$

No net work done. In this hypothetical reversible cycle heat has been transferred from a cold reservoir to a hot reservoir with no input of work. This cannot happen.

e) Assume an efficiency of 0.6. Then $-w''/q_1'' = 0.6$

For $w'' = +100\,\text{kJ}, q_1'' = -(100\,\text{kJ})/(0.60) = -167\,\text{kJ}; q_3'' = +67\,\text{kJ}$

Engine (a) could then be used to drive this one in reverse, as a refrigerator. As a result

$q_3 + q_3'' = -33 + 67 = 34\,\text{kJ}$, heat absorbed at T_{cold}

$q_1 + q_1'' = 133 - 167 = -34\,\text{kJ}$, heat discharged at T_{hot}

No net work done. Again, an impossible result occurs.

3. The organisms in the culture grow by using chemical energy from the nutrient medium. The gain in entropy from this conversion of foodstuffs more than compensates for the loss in entropy associated with the "growth" of the organism.

4. a) $\Delta S_{\text{mix}} = \Delta S_A + \Delta S_B = -n_A R \ln(P_A/1\,\text{atm}) - n_B R \ln(P_B/1\,\text{atm})$ from Eq. (3.21)

Now $n_A + n_B = 1$; therefore $X_A = n_A/(n_A + n_B) =$

$P_A/(1\ \text{atm})$ and $X_B = n_B/(n_A + n_B) = P_B/(1\ \text{atm})$

$\Delta \overline{S}_{\text{mix}} = \Delta S_{\text{mix}}/(n_A + n_B) = -X_A R \ln X_A - X_B R \ln X_B$

Q.E.D.

b) $\Delta \overline{S}_{\text{mix}} > 0$, because $X_A < 1$ and $\ln X_A < 0$, etc

A positive value for $\Delta \overline{S}_{\text{mix}}$ means that the process should occur spontaneously in an isolated system, which is what we expect for the mixing process.

c) At constant temperature, $\Delta \overline{G}_{\text{mix}} = \Delta \overline{H}_{\text{mix}} - T\Delta \overline{S}_{\text{mix}}$ (Eq. 3.32)

$\Delta \overline{H}_{\text{mix}}$ can be positive, negative or zero. For ideal gases or for the formation of an ideal liquid solution, $\Delta \overline{H}_{\text{mix}} = 0$. In this case $\Delta \overline{G}_{\text{mix}} = -T\Delta \overline{S}_{\text{mix}} < 0$, which is what we expect for the spontaneous mixing process.

5. a) The amount of heat released when $H_2O(g)$ condenses raises the temperature to $\underline{100°C}$.

b) $q_p = \Delta H = C_p(q)\Delta T + \Delta H\,(\text{condensation}) = 0$

(0.20 mol)(1.874 kJ K^{-1} kg^{-1})(0.018 kg mol^{-1})(5 K) + (Ù mol)(−40.66 kJ mol^{-1}) = 0

Ù = no. of mols of $H_2O(g)$ condensed = $\underline{8.3 \times 10^{-4}\ \text{mol}}$

c) $$\Delta S = C_p(\text{g})\ln\frac{T_2}{T_1} + \Delta S\,(\text{condensation})$$

$$= (0.20\,\text{mol})(1.874\,\text{kJ K}^{-1}\text{kg}^{-1})(0.018\,\text{kg mol}^{-1})\ln\frac{373}{368} + \frac{(8.3\times10^{-4}\,\text{mol})(-40.66\,\text{kJ mol}^{-1})}{373\,\text{K}}$$

$\Delta S = \underline{4.83\times10^{-7}\,\text{kJ K}^{-1}}$

6. a) The motor will not work. The ratchet and stick are also hit randomly by gas molecules. Thus, the stick, intended to only allow clockwise rotation of the wheel (ratchet), can be lifted due to collisions with gas molecules at the ratchet. At the particular moment when it is lifted, there is equal probability that the stick will slip forward or backward one notch due to the random fluctuations of the gas molecules at the paddle. The result is no net motion.

b) If the ratchet is at a lower temperature than the paddle, the likliehood that the stick will lift due to fluctuations at the ratchet will be much less (at lower temperature, sufficient energy for this process is less often achieved.) Thus, the lifting of the stick will be determined by the gas molecules hitting the paddle, and the stick will work as intended, only allowing clockwise rotation. In this scenario, the second law is not violated, i.e. heat is not converted to work at constant temperature.

7. a) For any cyclic process, all properties of state will remain unchanged. Thus, $\Delta T = \Delta E = \Delta H = \Delta S = \Delta G = 0$ Assuming that the isothermal expansion is done at T_{hot} and the isothermal compression at T_{cold}, then $w < 0$ (net work is done by the system) and $q > 0$ (net heat is absorbed).

b) $w = 0$, (fixed volume); $q = 0$, (thermally insulated)
$\Delta E = 0$, (internal energy is conserved)
$\Delta H = 0 = \Delta E \Delta (PV)$, (neglects thermal expansion)
$\Delta S > 0$, (spontaneous process in isolated system)

c) $\Delta V > 0$, (expansion); $w < 0$, (work done in reversible expansion)
$q = 0$, (adiabatic); $\Delta E = w < 0$, (First Law)
$\Delta E = C_v \Delta T < 0$ implies that $\Delta T < 0; \Delta H = C_P \Delta T < 0$
$\Delta S = q_{\text{rev}}/T = 0$

d) $q < 0$, (metabolism converts chemical energy into heat)
$\Delta T = 0$ (thermostat); $w \cong 0$ (depending on whether gases are produced or consumed)
$\Delta E \cong q < 0; \Delta H \cong \Delta E < 0$
$\Delta G < 0$ (spontaneous process at constant T, P)

8. a) $1\,\text{kg H}_2 = (1000\,\text{g})/(2.016\,\text{g mol}^{-1}) = 496.0\,\text{mol}$

b) $1\,\text{kg of } n\text{-octane} = (1000)(114.23) = 8.75\,\text{mol}$

$\Delta H_f^\circ(H_2O,1) = (-285.83)(496) = \underline{-141,780\text{ kJ}}$

$\Delta S^\circ = [\bar{S}^\circ(\text{H}_2\text{O},1) - \bar{S}^\circ(\text{H}_2,\text{g}) - 0.5\bar{S}^\circ(\text{O}_2,\text{g})](496\text{ mol})$

$= [69.91 - 130.57 - (0.5)(205.03)](496) = \underline{-80.94\text{ kJ K}^{-1}}$

$\Delta G_f^\circ = (-237.18)(496) = \underline{-117,650\text{ kJ}}$

$12.5\ \text{O}_2(\text{g}) + \text{C}_8\text{H}_{18}(\text{g}) \rightarrow 8\text{CO}_2(\text{g}) + 9\text{H}_2\text{O}(\text{g})$

$\Delta H^\circ = [8(-393.51) + 9(-285.83) - (-208.45)](8.75) = \underline{-48{,}230\text{ kJ}}$

$\Delta S^\circ = [8(213.64) + 9(69.91) - 466.73 - (12.5)(205.03)](8.75)$

$= \underline{-6.05\text{ kJ K}^{-1}}$

$\Delta G^\circ = [8(-394.36) + 9(-237.18) - 16.40](8.75) = \underline{-46{,}430\text{ kJ}}$

9. a) $6\text{C(s, graphite)} + 3\text{H}_2(\text{g}) \rightarrow \text{C}_6\text{H}_6(\text{g})$

$\Delta G_{298}^\circ = \Delta G_{f,298}^\circ(\text{benzene}) = \underline{129.66\text{ kJ mol}^{-1}}$

Don't buy. The reaction is highly non-spontaneous at 25°C and 1 atm. A catalyst can only change the rate of reaction, not the spontaneity.

b) $2\ \text{NO(g)} + \text{O}_2(\text{g}) \rightarrow 2\text{NO}_2(\text{g})$

$\Delta G_{298}^\circ = -2(86.55) + 0 + 2(51.31) = \underline{-70.48\text{ kJ mol}^{-1}}$

Yes, the reaction is spontaneous.

c) $3\text{CH}_4(\text{g}) + \text{NH}_3(\text{g}) + 3\text{O}_2(\text{g}) \rightarrow \text{CH}_3\text{CHNH}_2\text{COOH(s)} + 4\text{H}_2\text{O(l)}$

$\Delta G_{298}^\circ = -3(-50.72) - (-16.45) + 0 + (-370.24) + 4(-237.129)$

$= \underline{-1150.15\text{ kJ mol}^{-1}}$

Yes, this reaction is spontaneous.

10. a) irreversibly

b) the system plus the surroundings. In this case, the surroundings is a correct answer but the first answer is true for all spontaneous irreversible processes.

c) enthalpy change

d) greater than

11. $C_2H_5OH(l) \rightarrow C_2H_6(g) + 1/2O_2(g)$

$\Delta H^\circ_{298} = -(-276.98) + (-84.68) + 0 = \underline{192.30 \text{ kJ mol}^{-1}}$

$\Delta G^\circ_{298} = -(-174.14) + (-32.82) + 0 = \underline{141.32 \text{ kJ mol}^{-1}}$

$\Delta E^\circ_{298} = \Delta H^\circ_{298} - \Delta(PV) \cong \Delta H^\circ_{298} - (\Delta n_{gases})RT$

(assuming ideal gas behavior)

$= 192.30 \text{ kJ mol}^{-1} - (3/2)(8.314/1000)(298)$

$= \underline{188.6 \text{ kJ mol}^{-1}}$

To obtain ΔH^o at 500^oC and 1 atm we would need heat capacity values for each of the reactants and products over that temperature range. To be realistic, we should include the enthalpy of vaporization of ethanol, which is a gas at 500^oC and 1 atm.

$$\Delta H^\circ_{798} = \Delta H^\circ_{298} + \int_{298}^{798} (\Delta C_P) dT + \Delta H^0{}_{vap}(C_2H_5OH)$$

12. $CaCO_3(\text{calcite}) + CaCO_3(\text{aragonite})$

a) $\Delta G^\circ_{298} = -(-1128.76) + (-1127.71) = +1.05 \text{ kJ mol}^{-1}$

No. The positive ΔG° says that the process is non-spontaneous.

b) Because the density of aragonite is greater than that of calcite, an increase in pressure should favor the conversion to aragonite, according to Le Chatelier's Principle.

c) $\Delta G_{298}(P) - \Delta G^\circ_{298} = \Delta\overline{V}(P - 1\text{ atm})$, where $\Delta G_{298}(P) < 0$ (Eq. 3.44)

$P = -1{,}050 \text{ J mol}^{-1}[(100.09 \text{ g mol}^{-1})(1/2.930 - 1/2.710) \text{ cm}^3\text{g}^{-1}]^{-1} \times$

$(82.05 \text{ cm}^3 \text{ atm deg}^{-1}/8.314 \text{ J deg}^{-1}) + 1 \text{ atn}$

$= \underline{3700 \text{ atm}}$

d) $\Delta S^\circ_{298} = 88.70 - 92.88 = \underline{-4.18 \text{ J K}^{-1} \text{ mol}^{-1}}$

Because ΔS° is negative, that means that ΔG° will increase with increasing temperature, making the conversion even less favorable.

13. $CH_3COCOOH(l) \rightarrow CH_3CHO(g) + CO_2(g)$

a) $\Delta G^\circ_{298} = -(-463.38) + (-133.30) + (-394.359) = \underline{-64.28 \text{ kJ mol}^{-1}}$

b) Assuming that both CH_3CHO and CO_2 remain gases at 25°C and 100 atm.

$\Delta G = \Delta G^\circ + RT\ln([P_2(CH_3CHO)/P_1][P_2(CO_2)/P_1])$

$= -64.28 + (8.314/1000)(298)\ln(100 \times 100) = \underline{-41.46 \text{ kJ mol}^{-1}}$

14. a) $\Delta\overline{E}_{273} = \Delta\overline{H}_{273} - P(\overline{V}_{solid} - \overline{V}_{liquid})$

$= (18.016 \text{ g mol}^{-1})[(-333.4 \text{ J g}^{-1}) - (1 \text{ atm})(1.093 - 1.000) \text{ ml g}^{-1} \times (0.1013 \text{ J ml}^{-1} \text{ atm}^{-1})]$

$= \underline{-6.007 \text{ kJ mol}^{-1}}$

b) $\Delta H^\circ_{273} = (18.016 \text{ g mol}^{-1})(-333.4 \text{ J g}^{-1}) = \underline{-6.007 \text{ kJ mol}^{-1}}$

c) $$\Delta S_{273} = \Delta H^{\circ}_{273}/T = -22.00\,\mathrm{J\,K^{-1}\,mol^{-1}}$$
$$= S^{\circ}_{273}(\mathrm{H_2O,s}) - S^{\circ}_{273}(\mathrm{H_2O,l})$$
$$= 41.0 - 63.2$$
$$= \underline{-22.0\,\mathrm{J\,K^{-1}\,mol^{-1}}}$$

d) $\Delta G^{\circ}_{273} = \underline{0}$ (reversible process at const. T,P) (Eq. 3.31)

e) $q_P = \Delta H^{\circ}_{273} = \underline{-600\,\mathrm{kJ}}$ (heat is released)

f) $w = -(1\ \mathrm{atm})(1.093 - 1.000)\ \mathrm{ml\ g^{-1}} \times (0.1013\ \mathrm{J\ ml^{-1}\ atm^{-1}})(18.016\ \mathrm{g\ mol^{-1}})$
$= \underline{-.170\ \mathrm{J\ mol^{-1}}}$ (work is done by the system)

15. a) decrease
b) remain unchanged
c) zero
d) entropy change of the surroundings, entropy change of the universe
e) decrease
f) remain unchanged
g) remain unchanged
h) more negative than

16. a) $$\underline{w = -P_m(\Delta V_m)}$$
$$\Delta E = q_m + w; \Delta H = q_m$$
$$\Delta S = \frac{q_m}{T_m}; \Delta G = 0$$

b) $$\Delta H = q_m + (C_{p,\beta} - C_{p,\alpha})(T^* - T_m)$$
$$\Delta S = \frac{q_m}{T_m} + (C_{p,\beta} - C_{p,\alpha})\ln\frac{T^*}{T_m}$$

c) $$\Delta G = \underline{0}$$
$$\Delta H = \underline{638\,\mathrm{kJ\,mol^{-1}}}$$
$$\Delta S = \frac{638\,\mathrm{kJ\,mol^{-1}}}{343\,\mathrm{K}} = \underline{1.86\,\mathrm{kJ\,K^{-1}\,mol^{-1}}}$$

d) $$\Delta H = 638\,\mathrm{kJ\,mol^{-1}} - (8.37\,\mathrm{kJ\,mol^{-1}\,K^{-1}})(37 - 70)\,\mathrm{K}$$
$$\Delta H = \underline{914\ \mathrm{kJ\ mol^{-1}}}$$
$$\Delta S = 1.86\ \mathrm{kJ\ K^{-1}\ mol^{-1}} - (8.37\ \mathrm{kJ\ mol^{-1}\ K^{-1}})\ln\frac{310}{343}$$
$$\Delta S = \underline{2.71\ \mathrm{kJ\ K^{-1}\ mol^{-1}}}$$
$$\Delta G = \Delta H - T\Delta S = 914\ \mathrm{kJ\ mol^{-1}} - (310\ \mathrm{K})(2.71\ \mathrm{kJ\ K^{-1}\ mol^{-1}})$$
$$\Delta G = \underline{73.9\ \mathrm{kJ\ mol^{-1}}}$$

e) Increasing pressure favors the side with the smallest molar volume, therefore T_m will be increased by an increase in pressure and reactants will be favored.

f)

(from Eq. 3.35)

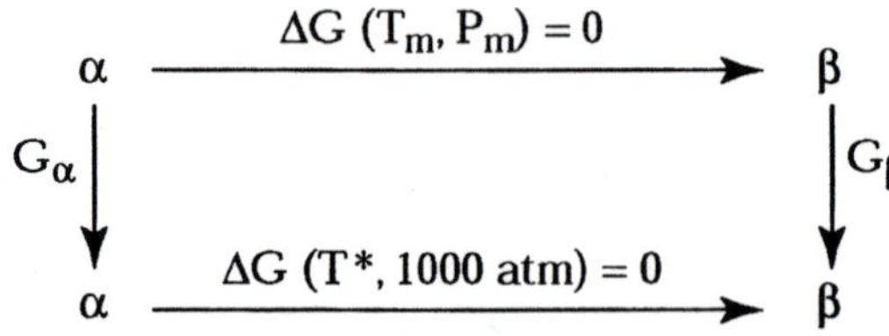

$$\alpha \xrightarrow{\Delta G\,(T_m, P_m) = 0} \beta$$

$G\alpha \downarrow \qquad \downarrow G\beta$

$$\alpha \xrightarrow{\Delta G\,(T^*, 1000\text{ atm}) = 0} \beta$$

$$\Delta G(T_m, P_m) = \Delta G(T^*, 1000 \text{ atm}) + G\alpha + G\beta$$

$$G_\alpha = \int_{P_m}^{1000} V_\alpha dP - \int_{T_m}^{T^*} S_\alpha dT = V_\alpha(1000 - P_m) - S_{\alpha,T_m} - C_{p,\alpha} \ln\frac{T^*}{T_m}$$

$$G_\beta = \int_{1000}^{P_m} V_\beta dP - \int_{T^*}^{T_m} S_\beta dT = V_\beta(P_m - 1000) - S_{\beta,T_m} - C_{p,\beta} \ln\frac{T_m}{T^*}$$

$$G_\alpha + G_\beta = 0 = (\Delta V)(P_m - 1000) - \Delta S(T_m) - \Delta C_p \ln\frac{T_m}{T^*}$$

$$\Delta V^\bullet = V_\beta - V_\alpha; \qquad \Delta C_p = C_{p,\beta} - C_{p,\alpha}$$

17. a) If more solvent is bound by the coil than the helix, both the entropy and enthalpy can decrease. Increasing the temperature favors the helix, not the coil.
Because ΔH is negative, K must decrease as T increases.

b) The reaction is spontaneous at 39°C.

$$\Delta G^\circ = \Delta H^\circ - T\Delta S^\circ$$

$$= -4.0 \text{ kJ} - (312\text{K})(-0.012 \text{ kJ K}^{-1})$$

$$\Delta G^\circ = -0.26 \text{ kJ}$$

$$T_m = \frac{\Delta H^\circ}{\Delta S^\circ} = \frac{-4.0 \text{ kJ}}{-0.012 \text{ kJ K}^{-1}} = 333 \text{ K} = \underline{60^\circ\text{C}}$$

d) No, for a spontaneous irreversible process $\Delta S_{univ} > 0$. If $\Delta S_{surr} = 0$ then $\Delta S_{sys} > 0$.

18. a) $H_2(g) + C\,(s, \text{graphite}) + O_2(g) \rightarrow HCOOH\,(g)$ All species are at 1 atm pressure

b) $\Delta H^\circ = -785.34 - 2(-362.63) = \underline{-60.08 \text{ kJ mol}^{-1}}$

$\Delta S^\bullet = 347.7 - 2(251.0) = \underline{-154.3 \text{ JK}^{-1} \text{ mol}^{-1}}$

$\Delta G^\bullet = -60.08 \text{ kJ mol}^{-1} - (298 \text{ K})(-0.1543 \text{ kJ K}^{-1} \text{ mol}^{-1})$

$\Delta G^\bullet = \underline{-14.1 \text{ kJ mol}^{-1}}$

c) $\Delta H^\bullet$ (hydrogen bond) $= -60.08/2 = \underline{-30.04 \text{ kJ mol}^{-1}}$

The entropy change and thus the free energy change include contributions from the loss of translational and rotational entropy. The values cannot be attributed simply to hydrogen bond formation

19. a) $\Delta G^\circ = \Delta G^\circ_f\,(NH_3) - \frac{1}{2}\Delta G^\circ_f\,(N_2) - \frac{3}{2}\Delta G^\circ_f\,(H_2)$

$\Delta G^\circ = \underline{-16.45 \text{ kJ mol}^{-1}}$

b) $\Delta G^\circ(T_2) = \Delta G^\circ(T_1) - (T_2 - T_1)\,(\Delta S^\circ)$

$\Delta S^\circ = 192.45 - (\frac{1}{2})(191.61) - (\frac{3}{2})(130.68)$

$\Delta S^\circ = -99.38 \text{ J K}^{-1}$

$\Delta G^\circ(50^\circ) = -16.45 \text{ kJ} - (25 \text{ K})(-0.09938 \text{ kJ K}^{-1}) = \underline{-13.96 \text{ kJ mol}^{-1}}$

Assume ΔS° is independent of T

c)) $100 \text{ g of } H_2 = (100 \text{ g})/(2.016 \text{ g mol}^{-1}) = 49.6 \text{ mol } H_2$

$$49.6 \text{ mol } H_2 = \frac{49.6 \text{ mol } H_2}{(\frac{3}{2} \text{ mol } H_2)/(\text{mol reaction})} = 33.07 \text{ mol reaction}$$

$\text{Free energy} = (33.07 \text{ mol})(-16.45 \text{ kJ mol}^{-1}) = -543.95 \text{ kJ}$

$$\text{Maximum time} = \frac{543.95\times10^3 \text{ J}}{10^{-3} \text{ J s}^{-1}} = \underline{5.44\times10^8 \text{ s}} \cong 17 \text{ yr}$$

20. a) An egg is an open system. Gases including O_2, CO_2 and H_2O pass through the pores of an egg shell. An isolated system has no interaction with the surroundings. No heat, work or matter is exchanged.

b) The entropy of a highly ordered chick is lower than the entropy of a solution of hen proteins. The entropy of the egg (the system) will depend on the mass gain or loss of the egg, and on its composition. The entropy of the system plus surroundings will increase in accordance` ` ` ` with the second law.

c) Matter and heat interchanges contribute to the energy change of the egg. The volume of the egg is essentially constant, so no work is done. The energy change could be calculated from the initial and final values for the amounts of each chemical in the egg.

d) The process is at constant T, P so the free energy decreases for a closed system. However, for an open system we must also know its change in mass and composition.

21. a) $\Delta S = n\overline{C}_P \ln (T_2/T_1)$ (Eq. 3.15)

$= (2 \text{ mol})(18.016 \text{ g mol}^{-1})(1.874 \text{ J K}^{-1}G^{-1}) \ln (373/393)$

$= \underline{-3.53 \text{ J K}^{-1}}$

b) $\Delta S = n\overline{C}_p \ln (T_2/T_1) = (1 \text{ mol})(8.314 \text{ J K}^{-1} \text{ mol}^{-1}) \ln (609/487)$

$= \underline{7.5 \text{ J K}^{-1}}$

c) $\Delta S = (100 \text{ g})(2.113 \text{ J K}^{-1} \text{ g}^{-1}) \ln (273/276) + (333.4 \text{ J g}^{-1}/273 \text{ K})(100 \text{ g}) + (100 \text{ g})(4.18 \text{ J K}^{-1} \text{ g}^{-1}) \ln (283/273)$

$= 7.88 + 122.12 + 15.05$

$= \underline{145.05 \text{ J K}^{-1}}$

22.

a) Supercooled, liquid water at $-10°C$ will spontaneously freeze to form ice and water at $0°C$. The entropy increase on warming the liquid from $-10°C$ to $0°C$ is balanced by the entropy decrease of freezing. The heat of freezing causes the temperature to rise. The fraction of water which freezes can be calculated from setting $\Delta H = 0$ for the process.

b) Using a heat pump is an efficient method of heating a house. It is always more efficient than directly converting work into heat (for example, by electrical heating). The maximum efficiency ratio for a heat pump is $T_{hot}/(T_{hot} - T_{cold})$; it is always greater than 1. This means each calorie of work done can pump more than 1 cal of heat into the building.

c) Do not do it. Refrigerators only transfer heat from inside the refrigerator to outside. If you turn off the refrigerator and open the door you can cool the room slightly.

d) The gases have spontaneously unmixed tending to decrease the entropy of the system, but because of the large increase in volume, the total entropy of the system has increased. As the system is isolated from the surroundings, there is no change in entropy of the surroundings. The entropy of the universe has increased and the second law is obeyed.

e) Wrong. The water will freeze, but the temperature will rise to give an increase in entropy. See discussion to part (a).

f) A catalyst allows a reaction to reach equilibrium. If H_2O_2 is decomposed on introducing the catalyst, it will not reform on removing the catalyst.

g) No. The highest temperature which could theoretically be obtained from a reflector is the sun's temperature. As the small object on which the sun is focused rises in temperature, it will emit radiation. At best it could approach temperature equilibrium with the sun.

h) The second law does apply to photosynthesis, but it is difficult to determine the effective temperature of the radiation from the sun. The effective temperature depends on the wavelengths emitted by the sun and absorbed by the plant. The high efficiency of energy conversion implies that the effective temperature of the absorbed radiation is about $2000°K$.

23. a) decrease
b) zero
c) decrease

24. a) temperature
b) temperature
c) increases
d) $\Delta P, \Delta T, \Delta G$
e) q, w. The properties of state, including ΔE, are all unaffected by the path. $w_{irrev} = 0; w_{rev} < 0.$

$$q_{rev} = \Delta E - w_{rev} > \Delta E = q_{irrev}$$

25. a) $dH = dE + PdV + VdP$ (from $H \equiv E + PV$)

$dE = dq + dw = dq - PdV$ (First Law)

$dq = TdS$ (Second Law)

therefore $dH = Tds + VdP$

b) $(\partial H/\partial P)_S = V$; $(\partial H/\partial S)_P = T$

c) $[\partial(\partial H/\partial P)_S/\partial S]_P = [\partial(\partial H/\partial S)_P/\partial P]_S$ (euler reciprocity relation)

therefore $(\partial V/\partial S)_P = (\partial T/\partial P)_S$

d) $dG = VdP - SdT$ (3.35)

$dG = dH - TdS - SdT$ (from $G \equiv H - TS$)

therefore $dG + SdT = VdP = dH - TdS$

therefore $dH = VdP + TdS$

26.
$$S_2 - S_1 = \Delta S = 0 = -R\ln(P_2/P_1) + \overline{C}_P \ln(T_2/T_1)$$
$$\ln(T_2/T_1) = (R/\overline{C}_P)\ln(P_2/P_1)$$
$$\ln T_2 = \ln(298) + (2/7)\ln(210/760) = 5.330$$
$$T_2 = 206\text{ K} = \underline{-67^\circ\text{C}}$$

27. The increase in pressure must provide a positive ΔG to counteract the negative ΔG associated with the spontaneous freezing at 253 K.

$$\Delta G_{253}(P) + \Delta G_{253}(1\text{ atm}) = 0$$
$$\int_1^P (\Delta V)DP + T_2[(\Delta G^\circ_{273}/273) + (1/T_2 - 1/273)\Delta H] = 0 \quad \text{(Eqs. 3.42a, 3.43a)}$$
$$(0.0903\text{ ml g}^{-1})(P - 1\text{ atm}) = -(253\text{ K})[0/273 + (1/253 - 1/273)\times(-333.4\text{ J g}^{-1})(9.869\text{ ml atm J}^{-1})]$$
$$P = 1\text{ atm} + 2669\text{ atm} = \underline{2670\text{ atm}}$$

CHAPTER 4

1. a) $K = \dfrac{[\text{G-1-P}][\text{ADP}]}{[\text{G}][\text{ATP}]} = (770)(10^{-4})/(10^{-3}) = \underline{77.0}$

$\Delta G^{\circ\prime}_{298} = -RT\ln K = -(8.314)(298)\ln 77 = \underline{-10.76\ \text{kJ mol}^{-1}}$

b)

		$\Delta G^{\circ\prime}$ $(kJ\ mol^{-1})$
$\text{G} + \text{ATP} \rightarrow \text{G}-1-\text{P} + \text{ADP}$	;	-10.76
$\text{ADP} + \text{P}_i \rightarrow \text{ATP} + \text{H}_2\text{O}$	;	$+31.10$
$\text{G} + \text{P}_i \rightarrow \text{G}-1-\text{P} + \text{H}_2\text{O}$		$\underline{20.2}$

$\ln K = -\Delta G^{\circ\prime}/RT = -(20{,}200)/(8.314)(298) = -8.17$

$K = \underline{2.9\times10^{-4}}$

2. a) $\text{ATP} \rightarrow \text{ADP} + \text{P}_i$

(Eq. 4.15)

$\Delta G' = \Delta G^{\circ\prime} + RT\ln([\text{ADP}][\text{P}_i]/[\text{ATP}])$

$= -31.00 + (8.314\times10^{-3})(298)\ln[(0.50\times10^{-3})(2.5\,x10^{-3})/(1.25\times10^{-3})]$

$= -31.0 - 17.1$

$= \underline{-48.1\ \text{kJ mol}^{-1}}$

b) $-w_{max,\,rev} = \Delta G' = \underline{48.1\ \text{kJ mol}^{-1}}$ (Eq. 4.55)

c)

		$\Delta G^{\circ\prime}$ (kJ mol^{-1})
$\text{PC} + \text{H}_2\text{O} \rightarrow \text{C} + \text{P}_i$	;	-16.7
$\text{P}_i + \text{ADP} \rightarrow \text{ATP} + \text{H}_2\text{O}$	;	31.0
$\text{PC} + \text{ADP} \rightarrow \text{C} + \text{ATP}$	;	-12.1

$\ln K = -\Delta G^{\circ\prime}/RT = 12{,}000/(8.314)(298) = 4.88$

$K = \underline{132}$

3. a)

		$\Delta G^{\circ\prime}$ (kJ mol^{-1})
$\text{G}-6-\text{P} + \text{ADP} \rightarrow \text{ATP} + \text{G}$	;	16.7
$\text{ATP} + \text{H}_2\text{O} \rightarrow \text{ADP} + \text{P}_i$	;	-31.0
$\text{G}-6-\text{P} + \text{H}_2\text{O} \rightarrow \text{G} + \text{P}_i$	;	$\underline{-14.3}$

b) $\ln K = -\Delta G^{\circ\prime}/RT = (16,700)/(8.314)(298) = 6.74$

$K = 846 = ([\mathrm{G-1-P}][\mathrm{ADP}])/([\mathrm{G}][\mathrm{ATP}])$

At equilibrium $[\mathrm{G-1-P}]/[\mathrm{G}] = \underline{846}$ when $[\mathrm{ADP}] = [\mathrm{ATP}]$

c) $\mathrm{G} + \mathrm{P}_i \rightarrow \mathrm{G-6-P} + \mathrm{H_2O}$

$\ln K = -\Delta G^{\circ\prime}/RT = -14,300/(298)(8.314) = -5.77$

$K = 3.11\times10^{-3} = [\mathrm{G\text{-}6\text{-}P}]/([\mathrm{G}][\mathrm{P}_i])$

$[\mathrm{G\text{-}6\text{-}P}]/[\mathrm{G}] = 3.11\times10^{-3}[\mathrm{P}_i] = \underline{3.11\times10^{-5}}$

4. a) $\mathrm{ATP} + 2\mathrm{G\,(out)} \rightarrow 2\mathrm{G\,(in)} + \mathrm{ADP} + \mathrm{P}_i; \Delta G^{\circ\prime} = -31.0\ \mathrm{kJ\ mol^{-1}}$

$\ln K = -\Delta G^{\circ\prime}/RT = 31,000/(8.314)(298) = 12.5$

$K = 2.72\times10^5 = [\mathrm{G(in)}]^2[\mathrm{ADP}][\mathrm{P}_i]/[\mathrm{G(out)}^2[\mathrm{ATP}]$

$[\mathrm{G(in)}]/[\mathrm{G(out)}] = \left[(2.72\times10^5)(10^{-2})/(10^{-2})^{-2}\right]^{1/2} = \underline{5,200}$

b) $[\mathrm{G\,(in)}]/[\mathrm{G\,(out)}] = \underline{2.72\times10^7}$

c) If $\gamma_{G(\mathrm{in})} < 1$, then $C_{G(\mathrm{in})} > 2.72\times10^7\,[\mathrm{G\,(out)}]$

This would increase the concentration gradient.

5.

	$\Delta G^{\circ\prime}_{303}$ (J mol^{-1})
Glu + Pyr $\longleftrightarrow$ Ket + Ala ;	−1004
Ket + Asp $\longleftrightarrow$ Glu + Ox ;	4812
Pyr + Asp $\longleftrightarrow$ Ala + Ox ;	3808

a) $K = [\mathrm{Ala}][\mathrm{Ox}]/[\mathrm{Pyr}][\mathrm{Asp}]$

$\ln K = -\Delta G^{\circ\prime}_{303}/RT = -3808/(8.314)(303) = -1.51$

$K = \underline{0.22}$

b) $\Delta G'_{303} = \Delta G^{\circ\prime}_{303} + RT\ln[(10^{-4})(10^{-5})/(10^{-2})(10^{-2})]$

$= 3808 + (8.314)(303)(-11.5) = \underline{-25,200\ \mathrm{J\ mol^{-1}}}$

The reaction proceeds in the forward direction under cytoplasmic conditions.

6. a) Least squares fit of lnK vs. 1/T gives: (Eq. 4.53)

$\ln K = -1.652 - 59.6(1/T)$

$\Delta H^{\circ} = -(\mathrm{slope})(R) = 59.6\,R = \underline{496\ \mathrm{J\ mol^{-1}}}$

b) at 25°C, $\ln K = -1.855$ and $K = 0.1564$

$\Delta G^\circ = -RT \ln K = \underline{4596 \text{ J mol}^{-1}}$

c) $$K = \frac{[2PG]}{[3PG]}$$

$$[2PG] + [3PG] = 0.150$$

$$[2PG] = \frac{0.150\text{ K}}{1+K} = \frac{(0.150)(0.1564)}{1.1564} = 0.020$$

$$[3PG] = \underline{0.130}$$

7. a) $\Delta H^\circ = \Delta H_f^\circ(SO_3) - \Delta H_f^\circ(SO_2) - (1/2)\Delta H_f^\circ(O_2)$

$= -395.7 + 296.8$

$\Delta H^\circ = -98.9 \text{ kJ mol}^{-1}$

$\Delta S^\circ = S^\circ(SO_3) - S^\circ(SO_2) - (1/2)S^\circ(O_2)$

$= 256.8 - 248.2 - (1/2)(205.1)$

$\Delta S^\circ = -93.95 \text{ J mol}^{-1}\text{ K}^{-1}$

$\Delta G^\circ = \Delta H^\circ - T\Delta S^\circ = -98.9\text{kJ mol}^{-1} - (298\text{ K})(-0.09395\text{kJ mol}^{-1}\text{ K}^{-1})$

$= \underline{-70.9 \text{ kJ mol}^{-1}}$

b) $K = e^{-\Delta G^\circ / RT} = e^{(70.9 \text{ kJ mol}^{-1}/(.008314 \text{ kJ mol}^{-1}\text{K}^{-1} \times 298\text{ K})}$

$K = 2.68 \times 10^{12}$

$$K = \frac{[SO_3]}{[0_2]^{1/2}[SO_2]}$$

$$\frac{[SO_3]}{[SO_2]} = K[O_2]^{1/2} = 2.68 \times 10^{12}(0.21)^{1/2} = \underline{1.23 \times 10^{12}}$$

c) $\Delta H^\circ = \Delta H_f^\circ(H_2SO_4) - \Delta H_f^\circ(H_2O) - \Delta H_f^\circ(SO_3)$

$$= -814.0 + 241.8 + 395.7$$

$$\Delta H^{\circ} = -176.5 \text{ kJ mol}^{-1}$$

$$\Delta S^{\circ} = 156.9 - 188.7 - 256.8$$

$$\Delta S^{\circ} = -288.6 \text{ J mol}^{-1} \text{ K}^{-1}$$

$$\Delta G^{\circ} = \Delta H^{\circ} - T\Delta S^{\circ} = -90.5 \text{ kJ mol}^{-1}$$

$$90.5/(0.008314)(298)$$

$$K = e^{(-\Delta G^{\circ}/RT)} = e^{90.5 \text{ kJ mol}^{-1}/(0.008314 \text{ kJ mol}^{-1}\text{K}^{-1} \times 298 \text{ K})}$$

$$K = 7.31 \times 10^{15}$$

$$K = \frac{[H_2SO_4]}{[H_2O][SO_3]}$$

$$\frac{[H_2SO_4]}{[SO_3]} = K[H_2O] = (7.31 \times 10^{15})(0.031) = \underline{2.26 \times 10^{14}}$$

8, a) $\Delta \overline{G}^{\circ} = -RT \ln K = -(8.314 \times 10^{-3})(298) \ \ln\ 4 = \underline{-3.43 \text{ kJ mol}^{-1}}$

b) $\Delta G = \underline{0}$ at equilibrium

c) $\Delta \overline{G} = \underline{-3.43 \text{ kJ mol}^{-1}}$

d) $\Delta G = 2(-3.43) = \underline{-6.86 \text{ kJ}}$

e) $\Delta \overline{H}^{\circ} = RT_i T_2 (\Delta T)^{-1} \ln(K_2/K_i)$ (Eq. 4.53)

$= (8.314 \times 10^{-3})(308)(298/10) \ln 2$

$= \underline{52.9 \text{ kJ mol}^{-1}}$

f) $\Delta \overline{S}^{\circ} = (\Delta \overline{H}^{\circ} - \Delta \overline{G}^{\circ})/T$

$= (52.9 + 3.4) \times 10^3/298$

$= \underline{189 \text{ J K}^{-1} \text{ mol}^{-1}}$

9. a) $\Delta G' = \Delta G^{\circ\prime} + RT \ln([ADP][P_i]/[ATP])$ (Eq. 4.15)

$= -31{,}000 + (8.314)(298) \ln[(10^{-4})(10^{-1})/(10^{-2})]$

$= -31{,}000 - 17{,}100 = \underline{-48.1 \text{ kJ mol}^{-1}}$

b) $-w_{\text{max, rev}} = -\Delta G' = \underline{48.1 \text{ kJ mol}^{-1}}$

c) $\Delta G' = -31{,}000 + (8.314)(298) \ln[(10^{-1})(0.25)/(10^{-7})]$

$= -31{,}000 + 30{,}800 = \underline{-200 \text{ J mol}^{-1}}$

$-w_{\text{max, rev}} = -\Delta G' = \underline{200 \text{ J mol}^{-1}}$

10. $CO(g) + 1/2O_2(g) \rightarrow CO_2(g); \Delta G^{\circ}_{298} = -394.359 + 137.168$

$$= -257 \text{ kJ mol}^{-1}$$

$$\ln K = -\Delta G^{\circ}/RT = 257{,}200/(8.314)(298) = 103.81$$

$$K = 1.2\times10^{45} = P_{CO_2}/P_{CO}P^{1/2}O_2$$

$$P_{CO} = (3\times10^{-4})/(0.2)^{1/2}(1.2\times10^{45}) = 5.5\times10^{-49} \text{ atm}$$

No need to worry about the *equilibrium* concentration.

11. a) $$\frac{d}{dT}\left(\frac{\mu_1^{\circ}}{T}\right) = \frac{d}{dT}\left(\frac{\overline{H}_1^{\circ}}{T}\right) - \frac{d}{dT}\left(\overline{S}_1^{\circ}\right) = -\frac{\overline{H}_1^{\circ}}{T^2} + \frac{1d\overline{H}_1^{\circ}}{TdT} - \frac{d\overline{S}_1^{\circ}}{dT}$$

but $\frac{d\overline{H}_1^{\circ}}{dT} = \overline{C}_p^{\circ}$ and $\frac{d\overline{S}^{\circ}}{dT} = \frac{\overline{C}_p^{\circ}}{T}$

$$\frac{d}{dT}\left(\frac{\mu_1^{\circ}}{T}\right) = -\frac{\overline{H}_1^{\circ}}{T^2}$$

b) $$\frac{d\ln K}{d(1/T)} = \left(\frac{d\ln K}{dT}\right)\left(\frac{dT}{d(1/T)}\right) = \left(\frac{\Delta\overline{H}^{\circ}}{RT^2}\right)\left(\frac{1}{-T^{-2}}\right) = \frac{-\Delta H^{\circ}}{R}$$

12. a) $\Delta\overline{G}^{\circ} = -RT\ln K = -(8.314)(298)(-10.96) = \underline{27.15 \text{ kJ mol}^{-1}}$

b) $\Delta\overline{G} = \underline{0}$ at equilibrium

c)

$$\Delta\overline{G} = \Delta\overline{G}^{\circ} + RT\ln[a_{HOAc}/a_{H+}a_{OAc^-}]$$

$$= -27{,}155 + (8.314)(298)\ln[1/(10^{-2})(10^{-4})]$$

$$= -27{,}155 + 34{,}230 = \underline{7.07 \text{ kJ mol}^{-1}}$$

d)

$$\Delta\overline{G} = -27{,}155 + (8.314)(298)\ln 10$$

$$= -27{,}155 + 5705 = \underline{-21.45 \text{ kJ mol}^{-1}}$$

e) $\Delta G = RT\ln(10^{-5}) = \underline{-28.5 \text{ kJ mol}^{-1}}$

13. a) $[Na^+] + [H^+] = [OH^-] + [H_2PO_4^{2-}] + 2[HPO_4^{2-}] + 3[PO_4^{3-}]$

b) $[HPO_4^{2-}]/[H_2PO_4^-] = K_2/[H^+] = (6.2\times10^{-8})/(10^{-7}) = 0.62$

Now $[H_3PO_4]$ and $[PO_4^{3-}] \ll [HPO_4^{2-}]$ or $[H_2PO_4^-]$ at pH 7

therefore $[HPO_4^{2-}] + [H_2PO_4^-) \simeq 0.100M = 0.62\ [H_2PO_4^-] + [H_2PO_4^-]$

$$[H_2PO_4^-] = 0.100\,M/1.62 = \underline{0.0617 M}$$

$$[HPO_4^{2-}] = 0.62[H_2PO_4^-] = \underline{0.0383\ M}$$

$$[H_3PO_4] = (10^{-7}[H_2PO_4^-]/(7.1\times10^{-3}) = \underline{8.7\times10^{-7} M}$$

$$[PO_4^{3-}] = [HPO_4^{2-}]K_3/[H^+]$$

$$= (0.0383)(4.5\times10^{-13})/(10^{-7}) = \underline{1.72\times10^{-7} M}$$

$$[Na^+] \simeq [H_2PO_4^-] + 2[HPO_4^{2-}] = \underline{0.1383\ M} \text{ from charge balance}$$

c) $$\ln K_2 = \ln K_1 - \Delta H^\circ (T_2^{-1} - T_1^{-1})/R \quad \text{(Table 4.2)}$$

$$= -16.60 - (4{,}150)(-1.299\times10^{-4})/8.314 = -16.54$$

$$K_2(310K) = \underline{6.58\times10^{-8}}$$

14. a) $pK_1 = 1.82;\ pK_2 = 6.00;\ pK_3 = 9.16$

at pH 9.0 HisH and His^- are principal species

$$K_3 = \frac{[His^-][H^+]}{HisH}$$

$$\frac{[His^-][H^+]}{HisH} = \frac{10^{-9.16}}{10^{-9.00}} = 0.6918$$

$$[His^-] + [HisH] = 0.200$$

$$0.6918[HisH] + [HisH] = \underline{0.200}$$

$$[HisH] = 0.200/1.6918 = \underline{0.118 M}$$

$$[His^-] = \underline{0.082\ M}$$

$$K_2 = \frac{[HisH][H^+]}{[His^+]} = 10^{-6.00}$$

$$[His^+] = \frac{(0.118)(10^9)}{(10^{-6})} = 1.18\times10^{-4}\ M$$

$$[His^{2+}] = \frac{(His^+)(H^+)}{(10^{-1.82})} = \frac{(1.18\times10^{-4})(10^{-9})}{(10^{-1.82})} = 7.80\times10^{-12}\ M$$

$$[OH^-] = Kw/10^{-9} = 10^{-5}\ M$$

$$[Na^+] = [His^-] + [OH^-] - [H^+] - [His^+] - 2[His^{2+}]$$

$$[Na^+] = \underline{0.082 M}$$

b) $\Delta H_1 = 0;\ \Delta H_2 = 29.9\ \text{kJ mol}^{-1};\ \Delta H_3 = 43.6\ \text{kJ mol}^{-1}$

$$\ln\left[\frac{K(40^\circ C)}{K(25^\circ C)}\right] = \frac{-\Delta H^\circ}{R}\left(\frac{1}{313} - \frac{1}{298}\right) = (0.01934)(\Delta H^\circ \text{ in kJ})$$

$$K_1 = 10^{-1.82};\ K_2 = 1.78\times10^{-6};\ K_3 = 2.32\times10^{-9.16}$$

$$\Delta H_w = 55.84\ \text{kJ mol}^{-1} \quad Kw = 2.94\times10^{-14}$$

$$K_3 = \frac{[\text{His}^-][\text{H}^+]}{\text{HisH}}$$ this equilibrium predominates at pH 9

$\text{Kw} = [\text{H}^+][\text{OH}^{-1}]$

$[\text{Na}^+] \cong [\text{His}^-]$ because other charged species are small

$[\text{His}^-] + [\text{HisH}] = 0.200$

$[\text{Na}^+] = 0.082$ M because no volume change

$[\text{His}^-] = 0.082$ M

$[\text{HisH}] = 0.118$ M

$[\text{H}^+] = (0.118/0.082) K_3 = 2.31 \times 10^{-9}$ M

pH = $\underline{8.64}$

c) $$[\text{His}^+] = \frac{(0.118)(2.31\times10^{-9})}{(1.78\times10^{-6})}$$

$[\text{His}^{2+}] = \underline{1.53\times10^{-4}\text{ M}}$

$$[\text{His}^{2+}] = \frac{(1.53\times10^{-4})(2.31\times10^{-9})}{(10^{-1.82})}$$

$[\text{His}^+] = \underline{2.33\times10^{-11}\text{ M}}$

$[\text{OH}^-] = (2.94\times10^{-4})/(2.31\times10^{-9}) = \underline{1.27\times10^{-5}\text{ M}}$

15. a) $\Delta G'_{298} = \Delta G^{0'}_{298} + RT\ln([\text{GDP}][\text{P}_i]/[\text{GTP}])$

$= -RT\ln K'_{298} + RT\ln[(5\times10^{-3})(15\times10^{-3})(15\times10^{-3})/(50\times10^{-3})]$

$= (8.314)(298)\ln[(1.5\times10^{-3})/(1.9\times10^{5})] = \underline{-46.2\text{ kJ mol}^{-1}}$

b) $\Delta(\Delta G') = RT\ln(1/2) = \underline{-1.7\text{ kJ mol}^{-1}}$

c) At equilibrium the reaction proceeds largely to the right. Thus
$[\text{GTP}] = x, [\text{GDP}] = 0.055 - x,\ [\text{P}_i] = 0.065 - x$ where x is small

$K'_{298} = 1.9\times10^{5} = (0.055 - x)(0.065 - x)/x \cong (0.055)(0.065)/x$

$[\text{GTP}] = x \cong (0.105)(0.115)/(1.9\times10^{5}) = \underline{1.88\times10^{-8}\text{M}}$

$[\text{GDP}] = \underline{0.055\text{ M}}$

$[P_i] = \underline{0.065\text{ M}}$

16. $K_{50} = (2.57\times10^{-6})/(9.97\times10^{-4}) = 2.58\times10^{-3}$

$K_{100} = (1.4\times10^{-4})/(8.6\times10^{-4}) = 0.1628$

$\Delta H^{\circ} = -R(T_2^{-1} - T_1^{-1})^{-1}\ln(K_{100}/K_{50})$

$= -(8.314)(-2410)(4.146) = \underline{83.05\text{ kJ mol}^{-1}}$

17 a) $K_1 = [\text{LOOP}]/[\text{SS}] = 0.86 = x/(10^{-3} - x)$

$$8.6\times10^{-4}-0.86x=x; x=[\text{LOOP}]=8.6\times10^{-4}/1.86=\underline{4.62\times10^{-4}\text{ M}}$$

$$10^{-3}-x=[\text{SS}]=\underline{5.38\times10^{-4}\text{ M}}$$

If the solution is ideal, increasing the concentration will have no effect on the fraction of hairpin loop.

b) $\Delta G^{\circ}_{310}=-RT\ln K_{310}=-(8.314)(310)\ln 0.51=\underline{1.74\text{ kJ mol}^{-1}}$

$\Delta H_{310}=-R(T_2^{-1}-T_1^{-1})^{1}\ln(K_{310}/K_{298})$

$=-(8.314)(-7{,}698)\ln(0.51/0.86)=\underline{-33.4\text{ kJ mol}^{-1}}$

$\Delta S^{\circ}_{310}=(\Delta H^{\circ}-\Delta G^{\circ})/T=(-33.4+1.74)/0.310=\underline{-102\text{ J K}^{-1}\text{ mol}^{-1}}$

c) $[\text{SS}]+[\text{H}]+2[\text{DS}]=0.100\text{ M}$

$[\text{DS}]=K_2([\text{SS}])^2; [\text{H}]=K_1[\text{SS}]$

$[\text{SS}]+0.86[\text{SS}]+2(10^{-2})([\text{SS}])^2=0.100\text{ M}$

$1.86[\text{SS}]+2\times10^{-2}[\text{SS}]^2=0.100\text{ M}$

$[\text{SS}]=\underline{0.0537\text{ M}}$

$[\text{H}]=\underline{0.0461\text{ M}}$

$[\text{DS}]=\underline{2.9\times10^{-5}\text{ M}}$

18) $Fe(CN)_6^{3-}+Cyt_{red}-Fe(CN)_6^{4-}+Cyt_{ox}$

a) At equilibrium (Eq. 4.61)

$$E^{\circ'}_{Cyt}-(RT/nF)\ln([Cyt_{red}]/[Cyt_{ox}])=E^{\circ'}_{FeCN}-(RT/nF)x$$
$$\ln([Fe(CN)_6^{4-}]/[Fe(CN)_6^{3-}])$$

$E^{\circ'}_{Cyt}=0.440-(0.0591)\log(2.0/0.1)=\underline{0.363\text{ volts}}$

b) $O_2+4H^{+}+2e^{-}\rightarrow 2H_2O$ $E^{o'}=.816$ volts at pH 7 (Table 4.1)

$E^{o}_{reaction}=.363-.616=-.415\text{ V}$

$\Delta G^{o}=-nFE^{o}>0$ so not spontaneous

Thus Cyt_{ox} is not a good enough oxidizing agent at pH 7.0 to cause the formation of O_2 from H_2O

19. Pyruvate $+2H^{+}+2e^{-}=$ Lactate; $E^{\circ\prime}=-0.18$ v

$\underline{\text{Cyt c}(\text{Fe}^{\text{III}}) + e^- = \text{Cyt}c(\text{Fe}^{\text{II}});\quad E^{\circ\prime} = +0.25\ \text{v}}$

a) $2\ \text{Cyt}\ c(\text{Fe}^{\text{II}}) + \text{pyr} + 2\text{H}^+ = 2\text{Cyt c}\ (\text{Fe}^{\text{III}}) + \text{lac};\ E^{\circ\prime} = \underline{-0.43\ \text{v}}$

b) $\log K' = E^{\circ\prime}/(0.0591/2) = -14.55$

$K' = \underline{2.8 \times 10^{-15}}$

c) $\Delta G^{\circ\prime} = -RT \ln K = -(8.314)(298)(-34.29) = \underline{83.9\ \text{kJ mol}^{-1}}$

d) $\Delta G' = \Delta G^{\circ\prime} + RT \ln([\text{lac}][\text{Cyt c}(\text{Fe}^{\text{III}})]^2/[\text{pyr}][\text{Cyt c}\ (\text{Fe}^{\text{II}})]^2)$

$= 83{,}900 + (0.008314)(298)\ \ln[5(10)^2] = \underline{99.2\ \text{kJ mol}^{-1}}$

---

20. a) $\text{Cl}_2(\text{g}) + 2\text{Ag(s)} + 2\text{I}^- + 2\text{AgI(s)} + 2\text{Cl}^-$

b) $\Delta G_{298} = -(96485)nE = -(96485)(2)(1.5702) = \underline{-303.0\ \text{kJ mol}^{-1}}$ (Eq. 4.57)

c) $\Delta G^\circ = \Delta G - RT \ln([\text{Cl}^-]^2/[\text{I}^-]^2\, P_{\text{Cl}_2})$

$= -303{,}001 - (8.314)(298) \ln[(10^{-3})^2/(10^{-2})^2]$

$= \underline{-291.6\ \text{kJ mol}^{-1}}$

d) $E^\circ = -\Delta G^\circ/96{,}485n = \underline{+1.5111\ \text{v}}$

$E^0_{\text{AgI}\to\text{Ag+I-}} = -1.5111 + 1.359 = -0.152$ volts]

e) $E^0_{\text{Ag}^+\to\text{Ag}} = +0.799$ volts

	E°(v)
$\text{Ag}^+ + \text{e}^- \rightarrow \text{Ag}$	.799
$\text{AgI} \rightarrow \text{Ag}^+ + \text{I}^-$	?
	−.152

$E^\circ_{\text{AgI}} = -.951\ \text{v}$

$\log K_{sp} = (-0.951)/0.0591 = -16.09$ (Eq. 4.63)

$K_{sp} = \underline{8 \times 10^{-17}}$

f) $\Delta S^0_{298} = (96{,}485)(2)(\Delta E/\Delta T) = \underline{-183.3\ \text{J K}^{-1}\ \text{mol}^{-1}}$

21. a) $K = [\text{H}^+][\text{Fd}_{\text{red}}]/([\text{H}_2]^{1/2}[\text{Fd}_{\text{ox}}])$

$= (10^{-7})(1/3)(2/3) = \underline{5 \times 10^{-8}}$

b) $E^{\circ\prime}$ refers to standard states at pH7, so H^+ is not included in the equilibrium expression

$K'(\text{pH7}) = [\text{Fd}_{\text{red}}]/([\text{H}_2]^{1/2}[\text{Fd}_{\text{ox}}]) = \underline{0.5}$

$E^{\circ\prime} = (RT/nF) \ln K' = 0.0591 \log_{10} 0.5 = -0.0178\ \text{V}$

$\text{Fd}_{\text{ox}} + e^- + \text{Fd}_{\text{red}};\qquad E^{\circ\prime} = -0.0178 - 0.421 = \underline{-0.439\ \text{V}}$

22. a) $ACC^- + 2H^+ + 2e^- \rightarrow \beta - HB^-;\ E^{o\prime} = -0.346\ V$

$\underline{O_2 + 4H^+ + 4e^- \rightarrow 2H_2O;}\quad \underline{E^{o\prime} = +0.816\ V}$

$\beta - HB^- + \frac{1}{2}O_2 \rightarrow AAC^- + H_2O;\ E^{o\prime} = 1.162\ V$

$\Delta G^{o\prime} = -96.485(2)(1.162) = \underline{-224.2\ kJ\ mol^{-1}}$

$\ln K' = -\Delta G^{o\prime}/RT = 90.5$

$K' = \underline{1.99 \times 10^{39}}$

b) $[ACC^-]/[\beta - HB^-] = K'(P_{O_2})^{1/2} = (4.9 \times 10^{42})(0.2)^{1/2}$

$= \underline{8.9 \times 10^{38}}$

23. a) $AA + 2H^+ + 2e^- \rightarrow ET;\ E^{o\prime} = -0.197\ V$

$\underline{O_2 + 4H^+ + 4e^- + 2H_2O;\ E^{o\prime} = 0.816\ V}$

$ET + 1/2O_2 \rightarrow AA + H_2O;\ E^{o\prime} = 1.013\ V$

$E^o = E^{o\prime} = \underline{1.013\ V}$ because the net reaction does not involve H^+ or OH^-

b) $\Delta G^o = -96.485(2)(1.013) = \underline{-195.5\ kJ\ mol^{-1}}$

c) $\ln K = -\Delta G^o/RT = 78.90$

$K = \underline{1.8 \times 10^{34}}$

d) $E = E^o - (0.0591/2)\log_{10}[AA][H_2O]/([ET][O_2]^{1/2})$

$= 1.013 - (0.02955)\log_{10}(1/[(0.1)(4)^{1/2}])$

$= \underline{0.992\ V}$

e) $\Delta G = -96.485(2)(0.992) = \underline{-191.5\ kJ\ mol^{-1}}$

24. a) $Mg(s)\,|\,Mg^{2+}\,||\,Mg^{2+}, 0.10\ M\ ATP\,|\,Mg\ (s)$

$$E = \frac{-RT}{nF}\ln\frac{a(Mg^{2+} \text{in ATP})}{a(Mg^{2+})}$$

Equal concentrations of Mg^{2+} are placed in two compartments of an electrochemical cell containing Mg electrodes. The EMF is a measure of the ratio of the activities of Mg^{2+} in the presence and absence of the ATP.

b) $$K = \frac{[ATP \cdot Mg^{2+}]}{[ATP][Mg^{2+}]}$$

The voltage of the cell in part 9a) gives us the activity of Mg^{2+} in an ATP solution. Assuming dilute solutions $a_I = [I]$. By knowing a(Mg^+ in ATP) we can solve for the equilibrium constant.

$[Mg^{2+}]$ from EMF

$[ATP \cdot Mg^{2+}] = [Mg^{2+}]$ total $- [Mg^{2+}]$

$[ATP] = [ATP]$ total $- [Mg^{2+}]$ total $+ [Mg^{2+}]$

25. a) $P700^{+} + e^{-} \rightarrow P700; \quad E^{o\prime} = 0.480$ V

$\underline{A + e^{-} \rightarrow A^{-}; \quad E^{o\prime} = -0.900}$

$P700 + A + P700^{+} \rightarrow A^{-}; \; E^{o\prime} = \underline{-1.380 \text{ V}}$

b) $\Delta G^{o\prime} = -96.485 E^{o\prime} = \underline{133.2 \text{ kJ mol}^{-1}}$

c) $NADP^{+} + H^{+} + 2e^{-} \rightarrow NADPH; \; E^{o} = -0.324$ V

$\underline{2H^{+} + 2e^{-} \rightarrow H_2(g); \; E^{o\prime} = -0.421 \text{ V}}$

$NADP^{+} + H_2(g) \rightarrow NADPH + H^{+}; \; E^{o\prime} = 0.097$ V

$\Delta G^{o\prime} = -(96.485)(2)(0.071) = \underline{-13.70 \text{ kJ mol}^{-1}}$

26. a) $E = E^{o} - (RT/nF)\ln[MB(red)]/([MB(ox)][H^{+}]^2)$

b) $E = 0.40 - (0.02955)\log_{10}[(1\times10^{-3})/(10^{-7})^2]$

$= \underline{0.075 \text{ V}}$

27. a) $NAD^{+} + H^{+} + 2e^{-} + NADH; \quad E^{o} = -0.320$ V

b) $\underline{1/2O_2 + 2H^{+} + 2e^{-} + 2H_2O; \quad E^{o\prime} = 0.816 \text{ V}}$

$NADH + H^{+} + 1/2O_2 + NAD^{+} + H_2O; \; E^{o\prime} = 1.136$ V

$\Delta G^{o\prime} = -96.485(2)(1.136) = -219.2 \text{ kJ mol}^{-1}$

$\Delta G' = \Delta G^{o\prime} + RT\ln([NAD^{+}]/[NADH][O_2]^{1/2})$

$= -219{,}214 + (8.314)(298)\ln[(2\times10^{-3})/(1\times10^{-3})(0.1)^{1/2}]$

$= \underline{-214.6 \text{ kJ mol}^{-1}}$

b) $\Delta G' = 3\Delta\overline{G}^{o\prime} + 3RT \ln [ATP]/([ADP][P_1])$

$= 3(31{,}000) + 3(8.314)(298)\ln[(3\times10^{-3})/(1\times10^{-3})(10\times10^{-3})]$

$= 93{,}000 + 42{,}400 = \underline{135.4 \text{ kJ mol}^{-1}}$

Fractional conversion to ATP $= 135.4/214.6 = \underline{0.63}$

28. a) $O_2 + 2\,\text{cysteine} + \text{cystine} + H_2O;\ E^{\circ\prime} = 1.16\ \text{V}$

$\log K = E^{\circ\prime}/(0.0591/2) = 1.16/0.02955 = 39.1$

$K = 1.3\times10^{39} = [\text{cystine}]/([\text{cysteine}]^2\,[O_2]^{1/2})$

$[\text{cystine}]/[\text{cysteine}]^2 = (1.73\times10^{39})(0.2)^{1/2} = 7.73\times10^{38}$

because reaction proceeds almost entirely to the right

$[\text{cystine}]_{eq} \cong .005\ \text{M}$

$[\text{cysteine}]_{eq} = [(0.005)/(7.73\times10^{38})]^{1/2} = 2.54\times10^{-21};$

$[\text{cystine}]/[\text{cysteine}] = \underline{1.96\times10^{18}}$

b) At equilibrium, $\underline{\Delta G = 0}$

29. $K_{obs} = K[\text{Na}^+]^{-\chi\psi}$

$\log K_{obs} = -\chi\psi \log[\text{Na}^+] +\ \text{constant}$

Each set of data falls on a good straight line

n = 4, slope = −3.68; $\chi_4 = 4.18$

n = 5, slope = −4.79; $\chi_5 = 5.45$

assuming $\psi = 0.88$

The ratio $\chi_5 / \chi_4 = 1.30$ which is close to 5/4.

This indicates that the length of the binding site is proportional to the number of lysines in the oligomers. Furthermore, each lysine appears to neutralize approximately 1 phosphate group.

30. $$K = \frac{[\text{D}]}{[\text{S}]^2}$$

f = fraction of single strands in double strand at equilibrium

$[\text{D}] = fc/2$

$2[\text{D}] + [\text{S}] = c$ (total concentration of single strands)

$[\text{S}] = C - 2[\text{D}] = c - fc = (1-f)c$

$$K = \frac{fc}{2(1-f)^2 c^2}$$

$$K = \frac{f}{2c(1-f)^2}$$

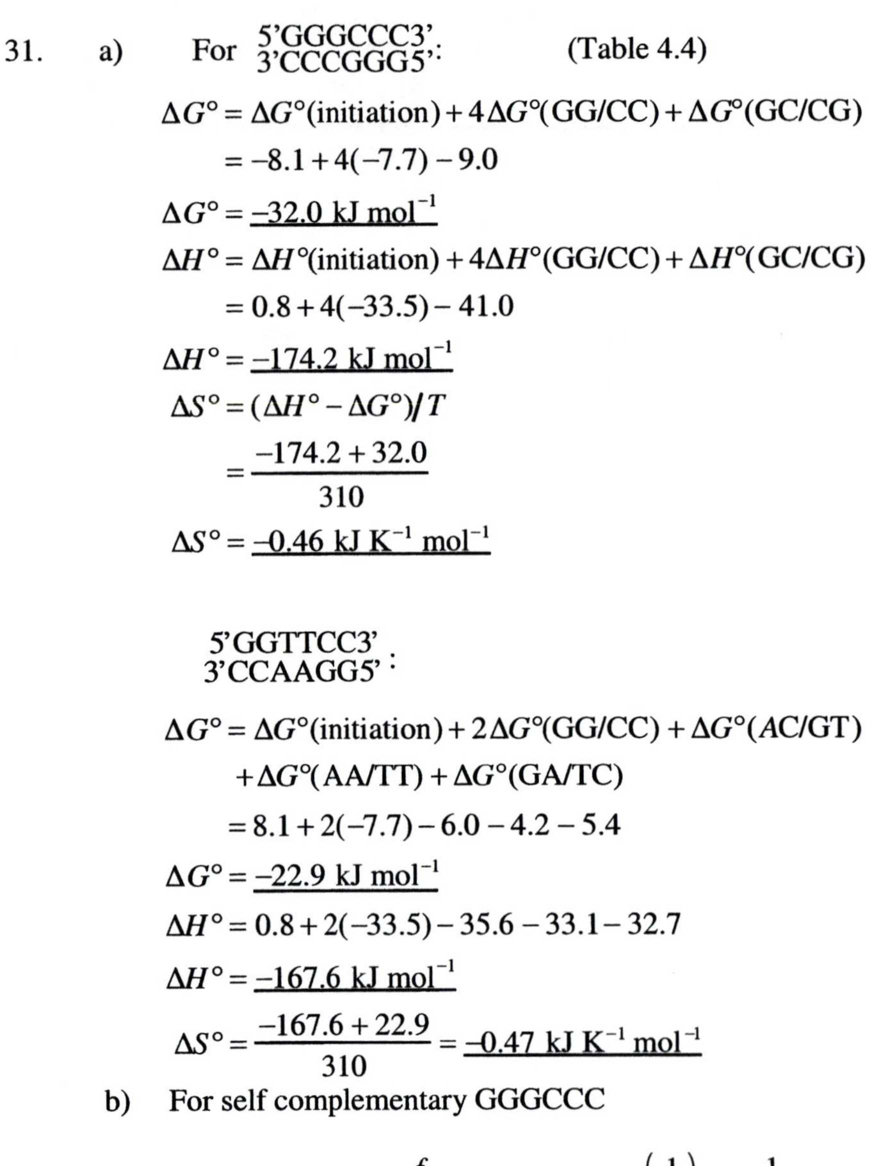

31. a) For $\begin{matrix}5'\text{GGGCCC}3' \\ 3'\text{CCCGGG}5'\end{matrix}$: (Table 4.4)

$$\Delta G^\circ = \Delta G^\circ(\text{initiation}) + 4\Delta G^\circ(\text{GG/CC}) + \Delta G^\circ(\text{GC/CG})$$
$$= -8.1 + 4(-7.7) - 9.0$$
$$\Delta G^\circ = \underline{-32.0 \text{ kJ mol}^{-1}}$$
$$\Delta H^\circ = \Delta H^\circ(\text{initiation}) + 4\Delta H^\circ(\text{GG/CC}) + \Delta H^\circ(\text{GC/CG})$$
$$= 0.8 + 4(-33.5) - 41.0$$
$$\Delta H^\circ = \underline{-174.2 \text{ kJ mol}^{-1}}$$
$$\Delta S^\circ = (\Delta H^\circ - \Delta G^\circ)/T$$
$$= \frac{-174.2 + 32.0}{310}$$
$$\Delta S^\circ = \underline{-0.46 \text{ kJ K}^{-1} \text{ mol}^{-1}}$$

$\begin{matrix}5'\text{GGTTCC}3' \\ 3'\text{CCAAGG}5'\end{matrix}$:

$$\Delta G^\circ = \Delta G^\circ(\text{initiation}) + 2\Delta G^\circ(\text{GG/CC}) + \Delta G^\circ(AC/\text{GT})$$
$$+ \Delta G^\circ(\text{AA/TT}) + \Delta G^\circ(\text{GA/TC})$$
$$= 8.1 + 2(-7.7) - 6.0 - 4.2 - 5.4$$
$$\Delta G^\circ = \underline{-22.9 \text{ kJ mol}^{-1}}$$
$$\Delta H^\circ = 0.8 + 2(-33.5) - 35.6 - 33.1 - 32.7$$
$$\Delta H^\circ = \underline{-167.6 \text{ kJ mol}^{-1}}$$
$$\Delta S^\circ = \frac{-167.6 + 22.9}{310} = \underline{-0.47 \text{ kJ K}^{-1} \text{ mol}^{-1}}$$

b) For self complementary GGGCCC

$$K = \frac{f}{2c(1-f)^2} \quad \text{at } T_m,\ f = \left(\frac{1}{2}\right)\ K = \frac{1}{c}$$
$$\Delta G^\circ = -RT_m \ln K = \Delta H^\circ - T_m \Delta S^\circ$$
$$T_m(\Delta S^\circ - R \ln K) = \Delta H^\circ$$
$$T_m = \frac{\Delta H^\circ}{\Delta S^\circ - R \ln K} = \frac{\Delta H^\circ}{\Delta S^\circ + R \ln c}$$
$$= \frac{-174.2 \times 10^3 \text{ J mol}^{-1}}{-460 \text{ J K}^{-1} \text{ mol}^{-1} + (8.314 \text{ J K}^{-1} \text{ mol}^{-1} \ln (1.0 \times 10^{-4}))}$$
$$T_m = 325 \text{ K } = 52^\circ\text{C}$$

For non-self-complementary GGTTCC + GGAACC

$$K = \frac{[\mathrm{D}]}{[\mathrm{S}_1][\mathrm{S}_2]}$$

$[D] = fc/2$ c = total concentration of single strands

$$[\mathrm{S}_1] = [\mathrm{S}_2] = \frac{c}{2}(1-f)$$

$$K = \frac{fc}{2\left[\frac{c}{2}(1-f)\right]^2} = \frac{2f}{c(1-f)^2}$$

$$K = \frac{4}{c} \text{ at } f = \frac{1}{2} \text{ at } T_m$$

$$\mathrm{Tm} = \frac{\Delta H°}{\Delta S° + R\ln(c/4)}$$

$$\mathrm{Tm} = \frac{\Delta H°}{\Delta S° + R\ln(c/4)} = \frac{-167.6\times10^3\,\mathrm{J\,mol^{-1}}}{-470\,\mathrm{J\,K^{-1}\,mol^{-1}} + 8.314\,\mathrm{J\,K^{-1}\,mol^{-1}}\ln\left(\frac{2\times10^{-4}}{4}\right)}$$

$$\mathrm{Tm} = 303\,\mathrm{K} = 30°\mathrm{C}$$

32. a)

$$K = \frac{[\mathrm{D}]}{[\mathrm{P}][\mathrm{T}]}$$

$$\mathrm{C_p} = [\mathrm{P}] + [\mathrm{D}]$$

$$\mathrm{C_T} = [\mathrm{D}] + [\mathrm{T}]$$

$$f = [\mathrm{D}]/C_T$$

$$\mathrm{C_p} \gg [\mathrm{D}] \quad \text{so} \quad [\mathrm{P}] = \mathrm{C_p}$$

$$[\mathrm{D}] = f\mathrm{C_T}$$

$$[T] = \mathrm{C_T} - [\mathrm{D}] = (1-f)\mathrm{C_T}$$

$$K = \frac{f\mathrm{C_T}}{\mathrm{C_p}(1-f)\mathrm{C_T}} = \frac{f}{(1-f)\mathrm{C_p}}$$

b)

$$\Delta G° = \Delta G°(\text{initiation}) + 3\Delta G°(\text{GG/CC}) + 2\Delta G°(\text{GA/CT}) + \Delta G°(\text{AA/TT}) + \Delta G°(\text{AT/TA}) + \Delta G°(\text{CA/TG})$$

$$= 8.1 + 3(-7.7) + 2(-5.4) - 4.2 - 3.7 - 6.0$$

$$\Delta G° = \underline{-39.7\ \mathrm{kJ\,mol^{-1}}}$$ (Table 4.4)

$$K = e^{-\Delta G°/RT} = \underline{9.0\times10^6}$$

c)

$$\Delta H° = 0.8 + 3(-33.5) - 34.3 - 33.1 - 30.2 - 32.7 - 35.6$$

$$\Delta H^{\circ} = -267.2 \text{ kJ mol}^{-1}$$

$$\Delta S^{\circ} = \frac{\Delta H^{\circ} - \Delta G^{\circ}}{T} = \frac{-267.2 + 39.7}{310} = -0.734 \text{ kJ K}^{-1} \text{ mol}^{-1}$$

$$T_m = \frac{\Delta H^{\circ}}{\Delta S^{\circ} - R \ln K} = \frac{\Delta H^{\circ}}{\Delta S^{\circ} - R \ln C_p} \quad \text{at } f = \frac{1}{2}$$

$$= \frac{-267.2 \times 10^3 \text{ J mol}^{-1}}{-734 \text{ J K}^{-1} \text{ mol}^{-1} + (8.314 \text{ J mol}^{-1}) \ln 10^{-4}}$$

$$T_m = 337\ K = \underline{64^{\circ}\text{C}}$$

d) As C_p increases T_m increases (see equation above)

As C_T increases T is unchanged: K is independent of target for large excess of probe.

A single base mismatch decreases ΔG° and ΔH° and decreases T_m

Lower salt concentration decreases stability of complex and decreases T_m

CHAPTER 5

1. a) $$\text{kg of water} = \left(\frac{4\,\text{J}}{\text{min cm}^2}\right)\left(\frac{100\,\text{cm}}{\text{m}}\right)^2\left(\frac{60\,\text{min}}{\text{hr}}\right)(8\,\text{hr})\left(\frac{\text{kg}}{2402\times10^3\,\text{J}}\right)$$
$$= \underline{7.99\,\text{kg}}$$

b) $$\text{Vapor pressure of } H_2O \text{ at } 40°\text{C} = (55.324\,\text{Torr})\left(\frac{\text{atm}}{760\,\text{Torr}}\right)$$
$$= 7.28\times10^{-2}\,\text{atm}$$
$$V = \frac{nRT}{P} = \frac{[(7990\,\text{g})/(18\,\text{g mol}^{-1})](0.08205\,\text{L atm K}^{-1}\,\text{mol}^{-1})(313\,\text{K})}{(7.28\times10^{-2}\,\text{atm})}$$
$$V = \underline{1.57\times10^5\,\text{L}}$$

c) Vapor pressure of H_2O at $20°\text{C} = 17.535$ Torr
$= 2.31\times10^{-2}$ at
$$n = \frac{VP}{RT} = \frac{(1.57\times10^5\,\text{L})(2.31\times10^{-2}\,\text{atm}}{(0.08205\,\text{L atm K}^{-1}\,\text{mol}^{-1})(293\,\text{K})}$$
$n = 151\,\text{mol}$
kg of water in air $= (151\,\text{mol})(18\times10^3\,\text{kg mol}^{-1}) = 2.72\,\text{kg}$
kg of water condensed $= 7.99 - 2.72 = 5.27$ kg
J of heat released $= (2447\,\text{kJ kg}^{-1})(5.27\,\text{kg}) = \underline{12895\,\text{kJ.}}$ The temperature of air increases.

d) $$\ln\frac{P_2}{P_1} = \frac{-\Delta H\,\text{vap}}{R}\left(\frac{1}{T_2}-\frac{1}{T_1}\right) \quad \text{(Eq. 5.3)}$$
$$\ln\left(\frac{P_2}{1}\right) = \frac{-(-40.66\,\text{kJ mol}^{-1})}{8.314\times10^{-3}\,\text{kJ K}^{-1}\text{mol}^{-1}}\left(\frac{1}{473\,\text{K}}-\frac{1}{373\,\text{K}}\right)$$
$$P_2 = \underline{15.9\,\text{atm}}$$

2. a) Vapor pressure of pyrene above a saturated solution of pyrene is P_1. At equilibrium the chemical potential of the pure pyrene is equal to the chemical potential of pyrene in solution and is equal to the chemical potential in the gas phase. If we choose the same standard state (the ideal gas at 1 atm) for the gas, solid and the solution, the activities are equal.
$a(\text{solid}) = a(\text{solution}) = a(\text{gas}) = P_1$. (Eq. 5.1)

b) Same as part (a). At equilibrium all activities are still equal to P_1.

c) At equilibrium for pyrene with the solute standard state:
$a(\text{solid}) = a(\text{solution}) = [P(\text{saturated solution})]$. $[P(\text{saturated solution})]$ = concentration of free pyrene in M in solution. It is independent of presence of cytosine.

$$K_1 = \frac{[P \cdot C(\text{soln})]}{[P(\text{soln})][C(\text{soln})]}$$

$$[P \cdot C(\text{soln})] = \text{solubility in cytosine soln} - \text{solubility in water}$$

$$= 1.1\times10^{-3} - 0.1\times10^{-3} = 1.0\times10^{-3}$$

$$[c(\text{soln})] = \text{conc of cytosine} - [P \cdot c(\text{soln})]$$

$$= 0.1 - 0.001 = 0.099 = 99\times10^{-3}$$

$$[P(\text{soln})] = \text{solubility in water} = 0.1\times10^{-3}$$

$$K_1 = \frac{(1\times10^{-3})}{(0.1\times10^{-3})(99\times10^{-3})} = \underline{101}$$

$$K_2 = \frac{[P \cdot C(\text{soln})]}{[C(\text{soln})]} = \frac{1.0\times10^{-3}}{99\times10^{-3}} = \underline{0.0101}$$

3. a) Under standard conditions (25°C, 1 atm) methane is a gas. It has low solubility in water and thus tremendous pressure would have to be applied to achieve the aqueous standard state. High pressure (low temperature) is also required for forming solid and liquid methane. This indicates that the gas must have the lowest standard free energy.

b)

$$P_B = k_B X_B \qquad X_B = 1 \text{ atm}/41\times10^3 \text{ atm} = 2.5\times10^{-5}$$

$$[CH_4]_{aq} = (2.5\times10^{-5})(55.6 \text{ M}) = \underline{1.35\times10^{-3} \text{ M}}$$

c) The solubility of CH_4 should be greater at 0^0C. The lower the temperature, the more likely the gas molecules will get "trapped" in solution.

d) As k increases with temperature, the mol fraction of the solute decreases (the concentration of dissolved gas decreases.) Thus $CH_4(g) \rightarrow CH_4(aq) + \text{heat}$. The enthalpy of dissolution is negative, the reaction is exothermic.

4. $\text{Myo} + O_2 \rightarrow \text{Oxymyo}$

$$K' = e^{-\Delta G^{\circ\prime}/RT} = 1.85\times10^5$$

$$K' = \frac{[\text{Oxymyo}]}{[\text{Myo}][0_2]}$$

$$[0_2] = 55.6X_{0_2}; X_{0_2} = P_{0_2}/k_{0_2}$$

$$k_{0_2} = 43\times10^3 \text{ atm from Table 5.3}; P_{0_2} = 30/760 \text{ atm}$$

$$[0_2] = 55.6\left(\frac{30}{760}\right)\left(\frac{1}{43\times10^3}\right) = 5.1\times10^{-5} \text{ M}$$

$$\frac{[\text{Oxymyo}]}{[\text{Myo}]} = K'[0_2] = (1.85\times10^5)(5.1\times10^{-5}) = \underline{9.4}$$

5. a) $\Delta G = \Delta G^\circ + RT \ln [\text{ATP}]/\text{ADP}][\text{P}_i]$

$= 31.0 + (8.314 \times 10^{-3})(298) \ln\ (10^{-3})/((10^{-3})(2.3 \times 10^{-3}))$

$= \Delta G = \underline{45.8 \text{ kJ mol}^{-1}}$

b) $\Delta G \quad = RT \ \ln \dfrac{a_{H^+}(\text{inside})}{a_{H^+}(\text{outside})} + \text{FZV}$

$= 2.303\ \text{RT}[\text{pH(outside)} - \text{pH(inside)}] + \ \text{FZV}$

$= 2.303(8.314 \times 10^{-3})(298)(-1.5) + (96.485(1)(-140 \times 10^{-3})$

$\Delta G = -22.1 \text{ kJ mol}^{-1}$ no, this is not enough to drive ATP synthesis

c) Moving slightly over two protons will provide enough energy to synthesize 1 ATP.

6. a) $\Delta \overline{G} = RT \ln (a_{\text{outside}} / a_{\text{inside}})$ (Eq. 5.25)

a = activity of NaCl (either inside or outside), but with the same standard state.

$\Delta \overline{G}$ = free energy change to transport 1 mol NaCl.

b) $\Delta \mu = RT \ln \dfrac{0.20}{0.05} = \underline{3573 \text{ J}}$. Not spontaneous.

c) $\Delta G = 3\Delta \overline{G}(\text{from b}) = \underline{10719 \text{ J}}$.

d) $\Delta \mu = 0$

e) $\Delta \mu = 0$

f) $\Delta G = \Delta G^\circ + RT \ln \dfrac{[\text{ADP}][\text{P}]}{[\text{ATP}]}$

$-40 = -31 + (10^{-3})(8.314)(310) \ln \dfrac{(1)[P]}{10}$

$\ln \dfrac{[\text{P}]}{10} = -3.49$

$[\text{P}] = \underline{0.30 \text{ M}}$

g) $K' = e^{-\Delta G^\circ / RT} = 1.67 \times 10^5 = \dfrac{[\text{ADP}][\text{P}]}{[\text{ATP}]}$

$[\text{P}] = \underline{1.67 \times 10^6 \text{ M}}$ It is clear that equilibrium cannot be attained.

7. Plot $v/[\text{A}]$ vs v. Within experimental error the data can be represented by a straight line with a slope of $1.0 \times 10^5 \text{ M}^{-1}$ and an intercept with the v-axis of 5. Thus, the data are in reasonable agreement with an identical and independent sites model. The intrinsic binding constant is $1.0 \times 10^5 \text{ M}^{-1}$, and there are about 5 sites per macromolecule.

8. a) From Eq. (5.16), we can see that for this case the v-axis intercept is 10, and the $(v/[\text{A}])$ axis intercept is 10 K or $5 \times 10^6 \text{ M}^{-1}$. A straight line through these two points is easily drawn.

b/c) We can solve Eq. 5.16 for υ to obtain

$$\upsilon = \frac{NK[\text{A}]}{1+K[\text{A}]}$$

For several classes of independent sites.

$$\nu_1 = N_1K_1[\text{A}]/(1+K_1[\text{A}])$$
$$\nu_2 = N_2K_2[\text{A}]/(1+K_2[\text{A}])$$

....

$$\nu_i = N_iK_i[\text{A}]/(1+K_i[\text{A}])$$

....

Adding all of the above equations, we obtain

$$\sum_i \nu_i = \sum_i \frac{N_iK_i[\text{A}]}{1+K_i[\text{A}]} = [\text{A}]\sum_i \frac{N_iK_i}{1+K_i[\text{A}]} \quad \text{or} \quad 1/[\text{A}] \sum_i \nu_i = \sum \frac{N_iK_i}{1+K_i[\text{A}]}$$

But $\sum_i \nu_i$ is just ν, the total number of bound A per macromolecules. Thus,

$$\frac{\nu}{[\text{A}]} = \sum_i \frac{N_iK_i}{1+K_i[\text{A}]}$$

This is the Scatchard equation for multiple classes of independent sites.
For the present case, $N_1 = 9,\ K_1 = 5\times10^5\ \text{M}^{-1};\ N_2 = 1,\ K_2 = 5\times10^6\ \text{M}^{-1}$. For any arbitrary [A], we can obtain $\nu/[\text{A}]$ from the equation above and then obtain ν. Some values are tabulated below.

	Part b		Part c	
[A]	$\nu/[\text{A}]$	ν	$\nu/[\text{A}]$	ν
10^{-8}	9.24×10^6	0.0924	2.63×10^7	0.263
10^{-7}	7.62×10^6	0.762	1.90×10^7	1.90
5×10^{-7}	5.03×10^6	2.51	9.14×10^6	4.57
10^{-6}	3.83×10^6	3.83	5.83×10^6	5.83
2×10^{-6}	2.70×10^6	5.41	3.52×10^6	7.05
4×10^{-6}	1.74×10^6	6.95	2.02×10^6	8.10
10^{-5}	8.48×10^5	8.48	9.07×10^5	9.07

Deviation from linearity is obvious only at low values of ν for part b; non-linearity is more apparent for part c. Experimentally, accurate values of $\nu/[\text{A}]$ are usually difficult to obtain at low values of ν.

9. a) $$\frac{v}{[\mathrm{A}]} = K \cdot (N - v)$$

$$N = 4,\ [\mathrm{A}] = 10^{-4}\,\mathrm{M},$$

$$[\text{Enzyme}] = \frac{\pi}{RT} = \frac{2.4\times10^{-3}}{(0.08205)(293)} = 1\times10^{-4}$$

$$v = \frac{2\times10^{-4}}{[\text{Enzyme}]} = 2$$

$$K = \frac{v}{[\mathrm{A}](N-v)} = \underline{10^{4}\,\mathrm{M}^{-1}}$$

$$\ln\frac{K_2}{K_1} = -\frac{\Delta H^\circ}{R}\left(\frac{1}{T_2} - \frac{1}{T_1}\right)$$

From y-intercepts of Scatchard plots $\frac{K_2}{K_1} = 2$ for $T_2 = 293$ K and $T_1 = 310$ K.

$$\Delta\overline{H}^\circ = \underline{38\ \mathrm{kJ}}$$

$$\Delta\overline{G}^\circ(293) = -RT\ln K = \underline{-22.4\ \mathrm{kJ}}$$

$$\Delta\overline{S}^\circ = \frac{\Delta H^0 - \Delta G^0}{T} = \underline{182\,\mathrm{J\ K^{-1}}}$$

10. a) $K_B(O_2) = 43\times10^3$ atm, $k_b(N_2) = 86\times10^3$ atm, $k_B(CO_2) = 0.05$ atm (Table 5.1, Eq. 5.11)

$$X_B = P_B / k_B$$

$c = 55.6X_B$ in dilute solution

$$c(O_2) = \frac{(55.6)(0.2)}{43\times10^3} = \underline{2.6\times10^{-4}\,\mathrm{M}}$$

$$c(N_2) = \frac{(55.6)(0.75)}{86\times10^3} = \underline{4.8\times10^{-4}\,\mathrm{M}}$$

$$c(CO_2) = \frac{(55.6)(0.05)}{1.6\times10^3} = \underline{17\times10^{-4}\,\mathrm{M}}$$

b) The total mole fraction of solutes is $X(O_2) + X(N_2) + X(CO_2)$

$$= \frac{(2.6 + 4.8 + 17)\times10^{-4}}{55.6} = 4.39\times10^{-5}$$

$$P_A = X_A P_o = (1 - X_B)P_o$$

$$P_A P_o = -X_B P_o = -(4.39\times10^{-5})(23.756) = -1.04\times10^{-3}$$

$$P_A = 23.756 - 0.001 = \underline{23.755\ \mathrm{torr}}$$

11. a) $\Delta T_b = K_b m = (0.51)(2) = 1.02$ (Eq. 5.52)

$T_b = \underline{101.02°C}$

b) Complex formation will lower the effective concentration of solute particles and lower the boiling point.

12. a) If the cells remain unchanged in 0.7% by weight NaCl, they have the same osmotic pressure as the salt. The freezing point gives the activity of water in the solution at the freezing point.

$$\ln a_{H_2O} = \frac{\Delta \overline{H}_{fus}}{R}\left(\frac{1}{T_o} - \frac{1}{T_f}\right) = \frac{\Delta \overline{H}_{fus}}{R}\left(\frac{T_f - T_o}{T_f T_o}\right) = -3.94 \times 10^{-3} \quad \text{(Eq. 5.53)}$$

Assume the a_{H_2O} is the same at 25°C. Then from Eq. (5.56)

$$\ln a_{H_2O} = \frac{\pi \overline{V}_{H_2O}}{RT} = -3.94 \times 10^{-3}$$

$$\pi = \frac{(3.94 \times 10^{-3})(82.05)(298)}{18.0}$$

$$\pi = \underline{5.35 \text{ atm}}$$

As a check we could assume ideal, dilute solution behavior. Then

$$m = \Delta T / K_f = 0.406 / 1.86 = 0.218 \text{ molal}$$

$$\pi = cRT \cong mRT = (0.218)(0.08205)(298) = 5.38\text{m}$$

b) concentration $= 0.22$ mol (kg solvent)$^{-1}$ from part (a)

$= 0.22$ mol (kg solvent)$^{-1}$(0.342 kg solute mol^{-1})

$= 0.0752$

$= \underline{7.5 \text{ wt\%}}$

13. a) $$\ln(P_2 / P_1) = \frac{-\Delta H_{vap}}{R}\left(\frac{1}{T_2} - \frac{1}{T_1}\right) \quad \text{(Eq. 5.3)}$$

$$\ln(60/760) = \frac{-(1368)(17)}{8.314}\left(\frac{1}{T_2} - \frac{1}{239.6}\right) = -2797\left(\frac{1}{T_2} - \frac{1}{239.6}\right)$$

$$T_2 = \underline{197 \text{ K}} = -76°\text{C}$$

b) $$\Delta S_{vap} = \Delta H_{vap} / T_b = \frac{(1368)(17)}{239.6} = \underline{97.1 \text{ J K}^{-1} \text{ mol}^{-1}}$$

Hydrogen bonding and association in liquids decreases the entropy of liquid, thus vaporization results in a greater than expected entropic gain.

c) $\Delta G° = -RT\ln K = -(8.314\times10^{-3})(223)\ln 10^{-30}$

$\Delta G° = \underline{128.1\text{ kJ}}$

The form of the equilibrium constant requires that a_{NH_3} for pure liquid ammonia be equal to 1.

d) First calculate the boiling point raising constant for NH_3. (Eq. 5.5)

$$K_b = \frac{M_{NH_3}RT_o^2}{1000\Delta H_{vap}} = \frac{(17)(8.314)(239.6)^2}{(1000)(1368)(17)} = 0.35$$

$$\Delta T = K_b m$$

$$m = \frac{33.4-32.7}{0.35} = \frac{0.70}{0.35} = 2$$

As the measured molality is 2 for a 1 molal solution of NH_4Cl in NH_3, the NH_4Cl must have completely dissociated.

14.

$$\frac{\Delta T}{\Delta P} = T\frac{\Delta V_{fus}}{\Delta H_{fus}} \quad \text{Eq. (5.37)}$$

$$\Delta T = -10,\ T = 268,\ \Delta V_{fus} = (1-1/0.9) = -0.111\text{ ml gm}^{-1}$$

$$\Delta H_{fus} = 333\text{ J gm}^{-1}$$

$$\Delta P(\text{atm}) = \frac{\Delta T\Delta H_{fus}}{T\Delta V_{fus}}\left(\frac{82.05\text{ ml atm}}{8.314\text{ J}}\right)$$

$$= \underline{1{,}110\text{ atm}}$$

15.

$$\ln\frac{P_2}{P_1} = \frac{-\Delta H_{vap}}{R}\left(\frac{1}{T_2}-\frac{1}{T_1}\right)$$

$$\ln\frac{P_2}{1} = \frac{-40.66}{8.314\times10^{-3}}\left(\frac{1}{393}-\frac{1}{373}\right)$$

$$P_2 = \underline{1.95\text{ atm}}$$

16. a) True.

b) False. The main effect of a pressure cooker is raising the boiling point of the water. The higher temperature achieved increases the rates of the cooking reactions.

c) False. The freezing point lowering constant depends on the properties of the solvent. (See equation 5.52). Equal molalities of different solutes in the same solvent will give the same freezing point.

d) True. Eq. 5.11 can be written as $X_B = (P_A^\circ - P_A)/P_A^\circ$.

e) False. Non-miscible liquids can be at equilibrium when small quantities of each dissolve in the other. The chemical potentials of each substance are equal in the different phases.

17. a)

$$X_B = (P_A^\circ - P_A)/P_A^\circ$$

$$X_B = \frac{n_B}{n_A + n_B} = \frac{\frac{54.66}{182.2}}{\frac{1000}{18} + \frac{54.66}{182.2}} = 5.37 \times 10^{-3}.$$

$$P_A^\circ - P_A = X_B P_A^\circ = (5.37 \times 10^{-3})(17.54) = \underline{0.0942 \text{ torr}}$$

b)

$$a_A = \frac{P_A}{P_A^\circ} = \frac{17.447}{17.54} = 0.9947 \quad \text{(Eq. 5.8)}$$

$$\gamma_A = \frac{a_A}{X_A} = \frac{0.9947}{1 - 5.37 \times 10^{-3}} = \frac{0.9947}{0.9946} = \underline{1.000}$$

c)

$$\pi = \frac{-RT \ln a_A}{\overline{V}_A} = \frac{-(82.05)(293) \ln 0.9947}{18} = \underline{7.1 \text{ atm}}$$

$$\pi \text{ ideal} = \left(\frac{54.66 \text{ g L}^{-1}}{182.2 \text{ g}}\right)(0.08205 \text{ L atm K}^{-1} \text{ mol}^{-1})(293 \text{ K}) = \underline{7.2 \text{ atm}}$$

18. a)

$$\frac{P_A^0 - P_A}{P_A^0} = \frac{\text{wt}_B}{M_B n_A} = \frac{1.00}{(180)\left(\frac{100}{18}\right)} = 10^{-3}$$

$$P_A^\circ - P_A = 17.54 \times 10^{-3}$$

$$P_A = \underline{17.52 \text{ torr}}$$

b)

$$\pi = cRT \quad c \cong \frac{1.00}{(180)(0.1)} = 0.0555 \quad \text{(Eq. 5.58a)}$$

$$\pi = (0.0555)(0.08205)(760)(293)$$

$$\pi = \underline{1015 \text{ torr}} \ (1.335 \text{ atm})$$

c)

$$\ln a_A = \frac{-\pi \overline{V}_A}{RT} = \frac{(1.335)(18)}{(82.05)(293)} = \underline{0.999}$$

d) Osmotic pressure gives a number average molecular weight, $\overline{M}_n$. Therefore Eq. (5.54) can be written as

$$\pi = wRT / \overline{M}_n w = \text{cong in } gl^{-1} \cong 20 \quad \text{(Eq. 5.59)}$$

$$\overline{M}_n = \frac{n_1 M_1 + n_2 M_2}{n_1 + n_2} = \frac{2}{\frac{1}{180} + \frac{1}{342}} = 236$$

$$\pi = \frac{(20)(0.08205)(760)(293)}{236} = \underline{1548 \text{ torr}}$$

19. From Henry's Law the mole fraction of N_2 dissolved is directly proportional to the pressure. The volume of N_2 gas dissolved, V_{N_2}, is also directly proportional to the pressure.

$$V_{N_2} = kP_{N_2}$$

$$k = \frac{1.3}{1} = 1.3 \text{ ml atm}^{-1} \text{ for } 100 \text{ ml blood.}$$

At 300 meters of water the air pressure in atm is approximately
$1 + [(1.025)(300)(1000)/(13.6)(760)] = 30.8$.

Therefore, $P_{N_2} = (0.78)(30.8) = 24.0$ atm.

$V_{N_2} = (1.3)(24.0) = 31.2$ ml N_2 dissolved for 100 ml blood at 300 meters.

Volume of N_2 liberated is

$$(31.2 - 1.0)\frac{(3200)}{100} = \underline{957 \text{ ml.}}$$

20. a) At equilibrium vapor pressure of water in aqueous solution is vapor pressure of pure water at 20°C.

$P_{H_2O} = \underline{17.535 \text{ torr}}$ from Table 2.2

$P_{C_6Cl_6} = \underline{1.00 \times 10^{-2} \text{ torr}}$ (same as pure solid)

b) (1) The vapor pressure of C_6Cl_6 is unchanged; it is determined by the solid C_6Cl_6.

(2) The vapor pressure of the water is decreased slightly because the protein decreases the activity of the water.

(3) The concentration of free C_6H_6 in the aqueous solution is unchanged; it is determined by the solid C_6Cl_6.

c) $$X_{C_6Cl_6} = \frac{P_{C_6Cl_6}}{P_{C_6Cl_6} + P_{H_2O}} = \frac{1.00 \times 10^{-2}}{1.00 \times 10^{-2} + 17.535}$$

$$X_{C_6Cl_6} = \underline{5.70 \times 10^{-4}}$$

d) $$\Delta H = \frac{-R\ln(P_2/P_1)}{(T_2^{-1}/T_1^{-1})} = \frac{-8.314 \text{ J K}^{-1}\text{ mol}^{-1}\ln(10^{-1}/10^{-2})}{[(313 \text{ K})^{-1} - (293 \text{ K})^{-1}]} = \underline{-87.8 \text{ kJ mol}^{-1}}$$

21. (a) $$\Delta\mu^\circ = -RT\ln\frac{[a_{G,aq}]}{[a_{G,s}]} = -(8.314 \times 10^{-3})(298)\ln(3.04)$$

$$\Delta\mu^\circ = \underline{-2.80 \text{ kJ mol}^{-1}}$$

b) $$\Delta\mu^\circ = -(8.314 \times 10^{-3})(298)\ln[(3.09)/(4.04 \times 10^{-4})]$$

$$\Delta\mu^\circ = \underline{-22.15 \text{ kJ mol}^{-1}} \text{ for Gly (ethanol)} \rightarrow \text{Gly}(H_2O)$$

c) $\Delta\mu^\circ$ (valine from water to ethanol) $= -RT\ln\dfrac{1.32\times10^{-3}}{0.60}$

$= +15.16 \text{ kJ mol}^{-1}$

$\Delta\mu^\circ$ (valine side chain from water to ethanol) $= 15.16 - (22.15) = \underline{-7.0 \text{ kJ mol}^{-1}}$

d) Yes, hydrophobic valine will have a lower chemical potential than glycine in the interior of the protein and will stabilize the folded state by 7 kJ mol^{-1}.

e)

$$\frac{K_{\text{val}}}{K_{\text{gly}}} = e^{-\Delta\mu^\circ/RT} = \exp\left[(7.0)/(8.314\times10^{-3})(298)\right]$$

$$\frac{K_{\text{val}}}{K_{\text{gly}}} = 16.9$$

The K for folding the protein with valine will be 16.9 times the K for folding the protein with glycine.

22. a) Assuming ideal behavior

$\Delta T_f = 1.86 \text{ m}$

$m = 0.56/1.86 = 0.30$ moles of solute per kg

Concentration of KCL $= \underline{0.15 \text{ molal}}$

b) $\pi = cRT = mRT = 2(0.15)(0.08205)(273) = \underline{6.72 \text{ atm}}$.

23. a) $\pi = cRT = (0.1)(0.08205)(300) = \underline{2.46 \text{ atm}}$.

Urea is assumed to be an ideal solution of unassociated urea molecules.

b) $\pi = cRT = \underline{4.92 \text{ atm}}$.

NaCl is assumed to be an ideal solution of completely dissociated ions.

c) The solvent will flow towards the side with the higher osmotic pressure. A pressure of 3 atm must be applied to prevent the flow.

d) $\Delta G = RT\ln(a_2/a_1)$

$$\ln a = \frac{-\pi\overline{V}_A}{RT}$$

$$\ln(a_2/a_1) = \ln a_1 = \frac{-(\pi_2 - \pi_1)\overline{V}_A}{RT}$$

$$\Delta G = -(\pi_2 - \pi_1)\overline{V}_A = -(2-5)(18) \text{ ml atm} = (3)(18)\left(\frac{8.314}{82.05}\right)$$

$\Delta G = \underline{5.47 \text{ J}}$

24. $M = wRT / \pi$

$$w = 6.0\ g\ l^{-1}; \pi = (22\ \text{mm of } H_2O)\left(\frac{1\ \text{mm Hg}}{13.65\ \text{mm } H_2O}\right)\left(\frac{1\ \text{atm}}{760\ \text{mm Hg}}\right)$$

$$\pi = 2.12 \times 10^{-3}\ \text{atm}$$

$$M = \frac{(6.0)(0.08205)(298)}{2.12 \times 10^{-3}}$$

$$M = \underline{69200}$$

25. $$\text{In}(P_2 / P_1) = \frac{-\Delta \overline{H}_{\text{vap}}}{R}\left(\frac{1}{T_2} - \frac{1}{T_1}\right)$$

$$P_2 = \underline{0.84\ \text{atm.}}$$

26. a) $zFV = -RTl\ (a_2 / a_1)$

$$V = \frac{(8.314)(310)}{(2)(96{,}487)} \ln \frac{0.1}{0.001} = \underline{-0.062\ \text{volts.}}$$

A negative potential of 0.062 volts must be present on the inside of the cell to keep the Ca^{++} in equilibrium.

b) $\Delta G = RTl\ (a_2 / a_1) + zFV$

$$= (8.314)(310)\ln \frac{0.1}{0.001} + (2)(96{,}487)(0.100) = \underline{31.2\ \text{kJ}}$$

$\Delta G = w_{\text{reversible}}$ if no pressure/volume work

27. a) $\pi = cRT \quad R = 0.08205\ l\text{K}^{-1}\ \text{mol}^{-1}, \quad T = 298$

Ocean water: $c = 1\text{M} \quad \pi = \underline{24.4\ \text{atm}}$

Lake water: $c = 0.015\ \text{M} \quad \pi = \underline{0.367\ \text{atm}}$

b) Use the following two equations (5.53 and 5.56):

$$\ln a = \frac{\Delta \overline{H}_{\text{fus}}}{R}\left(\frac{1}{T_o} - \frac{1}{T_f}\right) \text{ and } \ln a = \frac{-\pi \overline{V}_A}{RT}$$

$$T_o - T_f = \Delta T_{f\cdot p\cdot} = \frac{\pi \overline{V}_A T_o T_f}{\Delta \overline{H}_{\text{fus}} T} \cong \underline{\frac{\pi \overline{V}_A T_o}{\Delta \overline{H}_{\text{fus}}}}$$

c) $$\Delta G = RT \ln\left(\frac{{}^{a}H_2O^{\text{lake}}}{{}^{a}H_2O^{\text{ocean}}}\right)$$

$$= -(\pi\ \text{lake} - \pi\ \text{ocean})\ \overline{V}_A$$

$$= -(0.367 - 24.4)(18)\ \text{ml atm}\left(\frac{8.314\ \text{J}}{82.05\ \text{ml atm}}\right)$$

$$\Delta G = \underline{43.8\ \text{J}}$$

d) The lake has the highest vapor pressure of water.

e) $a = P/P^\circ = 747.7/760 = \underline{0.9838}$

28. a) ΔG is negative (activity of H_2O is lower in sucrose solution)
ΔH is approximately zero (ΔH dilution $= 0$)
ΔS is positive (ΔS mixing)
b) $\Delta G = 0$ (equilibrium)
ΔH is positive (vaporization)
ΔS is positive (vaporization)
c) ΔG is positive (equilibrium temperature is less than $0°C$)
ΔH is negative (freezing)
ΔS is negative (freezing)
d) ΔG is negative (equilibrium temperature is 100°C)
ΔH is negative (condensation)
ΔS is negative (condensation)

29. a) The concentration of pyruvate is .100 g/88 g mol^{-1}=.0011M
Assuming pyruvate completely dissociates;
$\pi = cRT = 2(.0011)RT = .053$ atm or $\underline{40.88 \text{ torr}}$
Assuming pyruvate does not dissociate;
$\pi = cRT = (.0011)RT = \underline{20.44 \text{ torr}}$
b) The extent of dissociation is directly proportional to the observed osmotic pressure.
40.88 torr (100% diss.) 20.44 torr (0% diss.) The difference is 20.44 torr.
36.9-29.44 torr = 16.46 torr due to dissociation.
% diss. = 16.46/20.44 = 81%

$$K_a = \frac{[H^+][Py^-]}{[Py]} = \frac{(.81 \times .0011M)^2}{(.0011 - 9 \times 10^{-4})} = .004$$

$pK_a = -\log K_a = \underline{2.4}$

30. a)

H_2O (liquid, 0°C) $\rightarrow$ H_2O (liquid, 100°C) $\Delta\bar{S} = \overline{Cp} \ln \frac{T_2}{T_1}$ = 23.47 J K^{-1} mol^{-1}

H_2O (solid, 0°C) $\rightarrow$ H_2O (liquid, 0°C) $\Delta\bar{S}$ = 22.22 J K^{-1} mol^{-1}

H_2O (liquid, 100°C) $\rightarrow$ H_2O (gas, 100°C) $\Delta\bar{S}$ = 109.2 J K^{-1} mol^{-1}

H_2O (gas, 1atm, 100°C) $\rightarrow$ H_2O (gas, .1 atm. 100°C) $\Delta\bar{S} \cong nR\ln\frac{P_2}{P_1}$ = 19.1 J K^{-1} mol^{-1}

b)

$$\Delta G_{tr} = RT \ln \frac{a_{outside}}{a_{inside}}$$

$\frac{.140}{.010} \rightarrow$ greatest ratio of activities so ΔG_{tr} largest for $\underline{\text{NaCl (.010 inside)} \rightarrow \text{NaCl(.140 outside)}}$

31. Water will evaporate from the 0.5 M sucrose solution and condense on the 1.0 M sucrose. The process will continue until both solutions have the same concentration of sucrose.

32. a) $$\ln a = \frac{-\pi\overline{V}}{RT} - \frac{-(25)(0.018)}{(0.08205)(273)} = -0.0201$$

$a = \underline{0.980}$

b) $P = aP^{\circ} = \underline{4.51 \text{ torr}}$

c) $$\ln a = \frac{\Delta H_{fus}}{R}\left(\frac{1}{T_o} - \frac{1}{T_f}\right)$$

$$\frac{1}{T_f} = \frac{1}{T_o} - \frac{R\ln a}{\Delta H_{fus}} = \frac{1}{273} - \frac{(8.314\times10^{-3})(-0.0201)}{5.86}$$

$T_f = \underline{271}$

33. The machine fails. It violates the second law by obtaining work at constant temperature in a cyclic process. The reason the machine does not work, however, is that gravity affects the distribution of gas molecules in the container. The vapor pressure is highest at the interface of the gas and liquid and decreases with increasing height.

The height difference between the two sides is h. On the pure solvent side, the barometric equation describes the fractional vapor pressure (at height h (P) vs. at the interface (P_o)):

$P = P_o \exp(-mgh/RT)$ where $m = .018 \text{ kg mol}^{-1}$ and $g = 9.8 \text{ m s}^{-2}$

$$\frac{P_o}{P} = \exp(-mgh/RT)$$

On the solution side of the membrane, the fractional vapor pressure (of the solution (P) vs. the pure solvent (P_o)) is related to the osmotic pressure by the following equations:

$$(1)\ a_A = \frac{P_o}{P} \quad (2)\ \ln a_A = \frac{-\pi\overline{V}_A}{RT} \quad \text{so} \quad \frac{P_o}{P} = \exp(-\pi\overline{V}_A/RT)$$

From the book we know that $\pi = \rho gh$ and $\pi\overline{V}_A = \rho gh\overline{V}_A$

$\rho\times\overline{V}_A = m$ (molar mass of water)

$$\text{so } \frac{P_o}{P} = \exp(-mgh/RT)$$

Notice that the expressions for the fractional pressure on each side are identical. Thus at height h, there is no pressure differential, and thus no vapor flow.

CHAPTER 6

1. a) From Eq. (6.16), the root-mean-square velocity is

$(u^2)^{1/2} = (3RT/M)^{1/2}$

$(3 \cdot 8.314 \text{ J deg}^{-1} \text{ mol}^{-1} \cdot 273 \text{deg}/2.016 \times 10^{-3} \text{kg mol}^{-1})^{1/2}$

$= \underline{1.838 \times 10^3 \text{ ms}^{-1}}$

b) From Eq. (6.14)

$\overline{U}_{tr} = \frac{3}{2}RT$

$= 1.5 \cdot 8.314 \times 10^{-3} \text{ kJ deg}^{-1} \text{ mol}^{-1} \cdot 273 \text{ deg}$

$= \underline{3.40 \text{ kJ mol}^{-1}}.$

c) According to the ideal gas law, the number of moles n per unit volume is

$/V = P/RT.$

The number of molecules per unit volume is

$n/V = N_o P/RT$

where N_o is Avogadro's number.

$N/V = 6.023 \times 10^{23} \text{mol}^{-1} \cdot 1 \text{ atm}/(82.05 \text{cm}^3 \text{ atm deg}^{-1} \text{ mol}^{-1} 273 \text{ deg})$

$= \underline{2.689 \times 10^{19} \text{ cm}^{-3}}.$

d) From Eq. (6.29) $l = 1/[\sqrt{2}\pi(N/V)\sigma^2]$

$N/V = 2.689 \times 10^{19} \text{cm}^{-3}$

$\sigma = 2.5 \times 10^{-8} \text{ cm}$

$l = \underline{1.34 \times 10^{-5} \text{ cm}}$

e) From Eq. (6.26)

$z = 4\sqrt{\pi} \cdot 2.689 \times 10^{25} \text{ m}^{-3} \cdot (2.5 \times 10^{-10} \text{ m})^2 (8.314 \text{ J})$

$\text{deg}^{-1} \text{ mol}^{-1} \cdot 273 \text{ deg}/2.016 \times 10^{-3} \text{ kg mol}^{-1})^{1/2}$

$= \underline{1.264 \times 10^{10} \text{ s}^{-1}}.$

f) From Eq. (6.27) $Z = (N/V)(z/2)$

$= 2.689 \times 10^{19} \text{ cm}^{-3} \cdot 1.264 \times 10^{10} \text{ s}^{-1}/2$

$= \underline{1.699 \times 10^{29} \text{ cm}^{-3}\text{s}^{-1}}.$

2. $\frac{n_2}{n_1} = e^{-(E_2 - E_1)/kT}$

a) For $n_2 = 0, \quad T \to 0\text{K}$

b) For $(n_2/n_1) = 1, \quad T \to \infty$

c) $\frac{n_2}{n_1} = e^{-h\upsilon/kT} = \frac{1}{1.000015}$

$\frac{h\upsilon}{kT} = 1.5 \times 10^{-5}$

$$T = \frac{(6.626\times10^{-34}\ \text{J s})(100\times10^{6}\ \text{s}^{-1})}{(1.5\times10^{-5})(1.38\times10^{-23}\ \text{JK}^{-1})}$$

$$T = 320\ K = 47^{\circ}\text{C}$$

3. a) From Eq. (6.43) $f = kT/D$

$$f = \frac{1.380\times10^{-6}\ \text{g cm}^2\ \text{s}^{-2}\ \text{deg}^{-1}\cdot 293\ \text{deg}}{4.0\times10^{-7}\ \text{cm}^2\ \text{s}^{-1}}$$

$$= \underline{1.01\times10^{-7}\ \text{g s}^{-1}}.$$

b) From Eq. (6.44)

$$f_o = 6\pi\eta\left(\frac{3M\bar{v}_2}{4\pi N_o}\right)^{1/3}$$

$\eta = 1.00\times10^{-2}\ \text{g s}^{-1}\ \text{cm}^{-1}$ for a dilute aqueous buffer at 20°C.

$$f_o = 6\pi\cdot 1.00\times10^{-2}\ \text{g s}^{-1}\ \text{cm}^{-1}\left(\frac{3\cdot 156{,}000\ \text{g mol}^{-1}\cdot 0.739\ \text{cm}^3}{4\pi\cdot 6.023\times10^{2}\ \text{mol}^{-1}}\right)^{1/3}$$

$$= \underline{6.74\times10^{-8}\text{g s}^{-1}}.$$

c) $f/f_o = 1.59$

If IgG is spherical, from Eqs. (6.44) and (6.47)

$$f/f_o = [(\bar{v}_2 + \delta_1\bar{v}_1^0)/\bar{v}]^{1/3}$$

$$\delta_1 = [(f/f_0)^3 - 1]\bar{v}_2/\bar{v}_1^0$$

$$= 2.22\ \text{cm}^3\ \text{g}^{-1}/v_1^0$$

We take $v_1^0 = 1\,\text{cm}^3\ \text{g}^{-1}$

$\delta_1 = \underline{2.22\ \text{g H}_2\text{O/g IgG}}$.

d) From Fig. 6.8 the axial ratio of a prolate corresponding to $f/f_o = 1.59$ is close to 10. The volume of an IgG molecule is $M\bar{v}_2/N_o$ or $1.91\times10^{-19}\ \text{cm}^3$.

$$\frac{4\pi}{3}ab^2 = \frac{4\pi}{3}\cdot 10b^3 = 1.91\times10^{-19}\text{cm}^{-3}.$$

$$b = 1.66\times10^{-7}\ \text{cm} = 16.6\ \text{Å}$$

$$a = 10b = \underline{166\text{Å}}$$

4. a) Many methods are described in the text.

b) Diffusion measurements provide the frictional coefficient of a protein. The radius of a spherical protein can be calculated. Changes in shape can be monitored,
Sedimentation velocity depends on molecular weight and frictional coefficient.
Combination of sedimentation and diffusion gives the molecular weight. Sedimentation equilibrium depends only on molecular weight (for an ideal solution)
Electrophoresis depends on net charge on protein and on frictional coefficient. Changes in charge with pH or mutations can be monitored. Gel electrophoresis in sodium

dodecylsulfate (SDS) can be calibrated to provide approximate, relative molecular weights.

c) The same molecular properties can be obtained for nucleic acids as for proteins. But for nucleic acids there is one negative charge per nucleotide phosphate. In a denaturing gel, electrophoresis is used to count nucleotides. This leads to a simple method of determining the sequence of a nucleic acid.

5. a) A plot of $\log x$ vs t gives a straight line with a slope of $3.32\times10^{-4}\ \text{min}^{-1}$ or $5.54\times10^{-6}\ \text{s}^{-1}$. From Eq. (6.58),

$$s=\frac{2.303}{\omega^2}\frac{d\log x}{dt}=\frac{2.303}{\omega^2}\cdot5.54\times10^{-6}\ \text{s}^{-1}$$

$$\omega=24{,}630\cdot2\pi/60\,\text{s}^{-1}$$

$$\omega^2=6.65\times10^6\ \text{s}^{-2}$$

$$s=2.303\cdot5.54\times10^{-6}/6.65\times10^{6}\text{s}$$

$$=1.96\times10^{-12}\ \text{s}=\underline{19.6\ \text{S}}.$$

b) From Eq. (6.52)

$$s_{20,\text{w}}=s\left[\frac{\eta}{\eta_{20,\text{w}}}\right]\left[\frac{(1-\bar{v}_2\rho)20,\text{w}}{1-\bar{v}_2\rho}\right]$$

$$=19.2\left(\frac{1.104}{1.005}\right)\left[\frac{(1-0.556\cdot0.998)}{(1-0.556\cdot1.04)}\right]\text{S}$$

$$=19.2\cdot1.098\cdot1.055\ \text{S}=\underline{22.3\ \text{S}}$$

6. a) From Eq. (6.53), the molecular weight of T7 phage is

$$M=\frac{RTs}{D(1-\bar{v}_2\rho)}$$

$$=\frac{8.314\times10^7\ \text{erg K}^{-1}\ \text{mol}^{-1}\cdot293\ \text{K}\cdot453\times10^{-3}\ \text{s}}{6.03\times10^{-8}\ \text{cm}^2\ \text{s}^{-1}(1-0.639\cdot0.998)}$$

$$=\underline{5.05\times10^7\,g\ \text{mol}^{-1}}.$$

b) The molecular weight of T7 DNA is
$5.05\times10^7\ \text{mol}^{-1}\cdot0.512=\underline{2.58\times10^7\ \text{g mol}^{-1}}$.

7. From Eq. (6.50),

$$s_1=\frac{m_1(1-\bar{v}_\rho)}{f_1}=\frac{M_1(1-\bar{v}_\rho)}{N_o f_1}$$

Similarly, $s_2=M_2(1-\bar{v}_\rho)/N_o f_2$.

Note that $\bar{v}$ is the same for the two viruses.

Thus $\dfrac{s_2}{s_1}=\dfrac{M_2}{M_1}\dfrac{f_1}{f_1}$.

Let r_1 and r_2 be the radii of the two spherical viruses.

$f_1/f_2 = r_1/r_2$

But

$$\frac{4\pi}{3}r_1^3 = \frac{M_1}{N_o}\bar{v}; \quad \frac{4\pi}{3}r_2^3 = \frac{M_2}{N_o}\bar{v}.$$

Thus $r_1/r_2 = (M_1/M_2)^{1/3}$

$$\frac{s_2}{s_1} = \frac{M_2}{M_1}\left(\frac{f_1}{f_2}\right) = \frac{M_2}{M_1}\left(\frac{r_1}{r_2}\right) = \frac{M_2}{M_1}\left(\frac{M_1}{M_2}\right)^{1/3} = \left(\frac{M_2}{M_1}\right)^{2/3}.$$

To obtain D_2/D_1, we resort to Eq. (6.93):

$D_2/D_1 = f_2/f_1 = (M_1/M_2)^{1/3} = (M_2/M_1)^{-1/3}$.

To obtain $[\eta]_2/[\eta]_1$, we note that unsolvated $(\delta_1 = 0)$ spherical particles of the same partial specific volumes have the same intrinsic viscosities [Eq. (6.65)]. Thus $[\eta]_2/[\eta]_1 = 1$.

8. The volume of a rod with length L and diameter d is $rd^2 L/4$.

The volume of a prolate ellipsoid of long and short semi-axes a and b is $4\pi ab^2/3$.

Equating these quantities, we obtain

$$\frac{\pi d^2 L}{4} = \frac{4\pi}{3}ab^2 = \frac{2\pi}{3}(2a)b^2.$$

But the length of the prolate, $2a$, is equal to L. Thus $d^2/4 = 2b^2/3$, and $d = (8/3)^{1/2}b$.

The ratio L/d for the rod is therefore

$$\frac{L}{d} = \frac{2a}{(8/3)^{1/2}b} = \left(\frac{3}{2}\right)^{1/2}\left(\frac{a}{b}\right).$$

9. a) $\bar{v} = \rho^{-1} = 0.8889 \text{ cm}^3\text{g}$

b) $s = v/g = 2 \text{ ms}^{-1}/9.81 \text{ ms}^{-2} = 0.204 \text{ s}^{-1}$

c) Eq. 6.65 $[\eta] = 2.5(\bar{v}_2 + \delta_1 v_1^0) = 2.5\frac{v_s}{m}$

$$v_s = \frac{[\eta]M}{(2.5)N_0}$$

$$v_s = \frac{[1.5 cm^3 g^{-1}]\cdot 5\times 10^6 g\cdot mol^{-1}}{(2.5)(6.02\times 10^{23})}$$

$$v_s = \underline{4.98\times 10^{-18} \text{ cm}^3}$$

d) Electrophoresis

e) Eq. 6.27 shows number of collisions is proportional to concentration squared, cross section squared, square root of temperature, and inversely proportional to square root to molecular weight. Answers are $\underline{4, 4, \sqrt{2}, 1/\sqrt{2}}$.

10. a) $$s = \frac{d \ln x}{\omega^2 dt}$$

$$\frac{d \ln x}{dt} = 3.42\times10^{-4}\ \text{min}^{-1} = (3.42\times10^{-4}\ \text{min}^{-1})\left(\frac{\text{min}}{60\ \text{sec}}\right)$$

$$= 5.70\times10^{-6}\ \text{sec}^{-1}$$

$$s = \frac{(5.70\times10^{-6}\ \text{sec}^{-1})(60\ \text{sec min}^{-1})^2}{(2\pi 18100\ \text{min}^{-1})^2}$$

$$s = 1.58\times10^{-12}\ \text{sec} = 15.85$$

$$M = \frac{RTs}{D(1-\bar{v}_2\rho)} = \frac{(8.314\times10^7\ \text{erg K}^{-1}\ \text{mol}^{-1})(293\ \text{K})(1.58\times10^{-12}\ \text{sec})}{(4.37\times10^{-7}\ \text{cm}^2\ \text{sec}^{-1})[1-(0.66)(1.02)]}$$

$$M = \underline{2.70\times10^5}$$

b) From gel electrophoresis pattern we see that the shortest fragment contains 200 base pairs.

c)
Molecular weight of protein = M(nucleosome) – M(DNA) = 270000 – (200)(660) = 138000

$$\text{Molecular weight of single protein} = \frac{138000}{8} = \underline{17250}$$

11. a) $$f = kT/D = \underline{3.6\times10^{-8}\ \text{gs}^{-1}}$$

$$f_o = 6\pi(0.010)\left[\frac{(3)(14{,}314)(0.703)}{4\pi(6.023\times10^{23})}\right]^{1/3}$$

$$f_o = \underline{2.12\times10^{-8}\ \text{gs}^{-1}}$$

b) $$\frac{s_2}{s_1} = \left(\frac{M_2}{M_1}\right)\left(\frac{f_1}{f_2}\right) = (2)(1.628) = 3.257$$

$$s_{20,w}\text{dimer} = 3.257(1.91\times10^{-13}) = \underline{6.22\times10^{-13}\ \text{sec}}$$

12. a) $$\frac{s_2}{s_1} = \left(\frac{M_2}{M_1}\right)\left(\frac{f_1}{f_2}\right) = \left(\frac{M_2}{M_1}\right)\left(\frac{r_1}{r_2}\right)$$

$$\frac{r_1}{r_2} = \left(\frac{M_1}{M_2}\right)^{1/3}$$

$$\frac{s_2}{s_1} = \left(\frac{M_2}{M_1}\right)^{2/3} = \left(\frac{200{,}000}{20{,}000}\right)^{2/3} = \underline{4.64}$$

b) $$\ln\frac{X}{X_o} = \omega^2 s(t - t_o)$$

$$\text{At } t_o = 0, X_o = 4 \text{ cm}$$

$$\frac{\ln(X_2/X_o)}{\ln(X_o/X_o)} = \frac{s_2}{s_1}$$

$$\ln(X_1/4) = \left(\frac{1}{4.64}\right)\ln(7/4)$$

$$X_1 = 4.5 \text{ cm}$$

The small protein has travelled $4.5 - 4 = \underline{0.5 \text{ cm}}$. This is a good way to separate the proteins.

c) $\langle X^2\rangle^{1/2} = (2Dt)^{1/2}$

$$D = kT/f = kT/(6\pi\eta r)$$

$$r = \left(\frac{3}{4\pi}\frac{M}{N_o}\bar{v}\right)^{1/3}$$

$$M_2 = 200{,}000;\ r_2 = 3.89\times10^{-7} \text{ cm};\ f_2 = 1.099\times10^{-7} \text{ g s}^{-1};$$

$$D_2 = 3.74\times10^{-7} \text{ cm}^2\text{s}^{-1}$$

$$\langle X_2^2\rangle^{1/2} = [2(3.74\times10^{-7})(4.32\times10^4)]^{1/2}$$

$$= \underline{0.18 \text{ cm}}$$

$$M_1 = 20{,}000; D_1/D_2 = f_2/f_1 = (M_2/M_1)^{1/3} = 2.154$$

$$D_1 = 8.06\times10^{-7} \text{ cm}^2\text{s}^{-1}$$

$$\langle X_1^2\rangle^{1/2} = \underline{0.26 \text{ cm}}$$

Diffusion will not interfere with separation.

d) No. After 14 hours large protein is at bottom of cell and $\langle X^2\rangle^{1/2} = 0.19$ cm

13. a) $t = \dfrac{\langle X^2\rangle}{2D}$

$$D = kT/6\pi\eta r = (1.381\times10^{-16})(298)/(6\pi)(0.10)(2\times10^{-8})$$

$$D = 1.09\times10^{-6}$$

$$t = \frac{(40\times10^{-8})^2}{2(1.09\times10^{-6})} = \underline{7.3\times10^{-8} \text{ s}}$$

b) The driving force for diffusion is the gradient of the chemical potential. Polar molecules in bilayers have a high chemical potential, so they diffuse slowly into the bilayer.

14. a) The SDS-gel gives a measure of the molecular weights of the mutants; mutant II has a lower molecular weight. The isoelectric focusing shows that mutant I has a lower isoelectric point (is more negatively charged) than the normal enzyme, but mutant II has a higher isoelectric point (is more positively charged). Mutant II must have lost some carboxyl groups, such as aspartic acid and glutamic acid. Mutant I must have exchanged a neutral side chain for a negative one, or a positive side chain for a neutral or negative one. Examples are lysine to alanine, or glycine to glutamic acid.

    b) The loss of a few amino acid residues from a linear peptide may not significantly change the apparent molecular weight as determined by SDS-gel electrophoresis. The shift in isoelectric point from pH 10 to pH 8 indicates the loss of some basic amino acids (pK = 11) such as lysine. A cyclic peptide will not change molecular weight when a single peptide bond is cleaved.

15. It is reasonable to assume that the partial specific volume $\bar{v}_2$ and the hydration parameter δ_1 for the dimer are the same as the corresponding quantities for the monomer. Then, from Eq. (6.65), $[\eta]_{\text{dimer}}/[\eta]_{\text{monomer}} = \nu_{\text{dimer}}/\nu_{\text{monomer}}$.
For a rod-like macromolecule, the length to diameter ratio L/d is proportional to the axial ratio (a/b) of a prolate ellipsoid with hydrodynamic properties similar to those of the rod (Problem 6.8). If the macromolecule forms an end-to-end dimer, L/d, and hence a/b of the corresponding prolate, doubles. Hence

$$\nu_{\text{dimer}}/\nu_{\text{monomer}} = 2^{1.732} = \underline{3.32}.$$

16. We can measure the diffusion coefficient before and after the digestion; the ratio s/D will not change if the molecular weight remains constant.

17. From Eq. (6.53) $M = \dfrac{RTs}{D(1-\bar{v}\rho)} = \underline{68{,}400}$

18. From Eq. 6.50 and Eq. 6.44

$$s = \frac{M}{f} = \frac{M}{r}$$

But Eq. 6.47 shows $r = [(\bar{v}_2 + \delta_1 v_1^0)]^{1/3}$

$$\text{Therefore } s = \frac{M^{2/3}}{(\bar{v}_2 + \delta_1 v_1^{\,0})^{1/3}}$$

$$\text{And } [\eta] = (\bar{v}_2 + \delta_1 v_1^{\,0})$$

$$\text{Therefore } s = \frac{M^{2/3}}{[\eta]^{1/3}}$$

Cleaving SV40 does not change the molecular weight, thus $s[\eta]^{1/3}$ for SV40 is independent of cleavage.

$$s_R[\eta]_R^{1/3} = s_L[\eta]_L^{1/3}$$
$$[\eta]_R/[\eta]_R = (s_L/s_R)^3 = (1/1.45)^3 = \underline{0.328}$$

19. If unfolding does not occur, $[\eta]$ should be insensitive to the molecular weight. If complete unfolding occurs, we expect that $[\eta]$ should be approximately proportional to the 1/2th power of the molecular weight. From the data given we can plot log $[\eta]$ in 6M guanidine $\cdot$ HCl vs log M. The slope of the log $[\eta]$ vs log M plot is 0.70, showing that unfolding occurs in 6M guanidine $\cdot$ HCL. The deviation of the slope from 0.5 can be attributed to the non-ideality of the system.

20. Table 4.1 gives the pKa values of the amino acids tyrosine and histidine. We are interested in the ionization of these residues in hemoglobin, but as an approximation we assume that the pKa values of a side chain of a residue in a protein are the same as those of the side chain of the amino acid.
For the histidine side chain, there is an ionizable proton with a pK_2 of 6.00:

$$\mathrm{H^+His} \rightleftharpoons \mathrm{H^+} + \mathrm{His}$$
$$\frac{[\mathrm{H^+His}]}{[\mathrm{His}]} = \frac{[\mathrm{H^+}]}{10^{-6}}$$
At $\mathrm{pH} = 7$,
$$\frac{[\mathrm{H^+His}]}{[\mathrm{His}]} = 0.1.$$
Thus the histidyl side chain is slightly positively charged.
The tyrosine side chain has no ionizable proton with a pK less than 10. At $\mathrm{pH} = 7$, it is uncharged. Therefore at pH 7 hemoglobin M Boston is expected to be slightly less positively charged than hemoglobin A. A reasonable pH for a trial separation of the two proteins would be 5. At this pH the normal hemoglobin A has about 2 positive charges more than the mutant hemoglobin M Boston.

21. a)
$$f = 6\pi\eta r = \frac{kT}{D}$$
$$r = \frac{kT}{6\pi\eta D} = \frac{(1.3807\times10^{-16}\,\mathrm{orgmol^{-1}K^{-1}})(313\,\mathrm{K})}{6\pi(0.0101\,\mathrm{gcm^{-1}s^{-1}})(14.25\times10^{-7}\,\mathrm{cm^2s^{-1}})}$$
$$r = 1.59\times10^{-7}\,\mathrm{cm} = 15.9\ \text{Å}$$
$$d = 2(15.9\ \text{Å}) = \underline{31.8\ \text{Å}}$$

b)
$$v(\mathrm{pH7}) = \frac{4}{3}\pi r^3 = 16.8\times10^3\ \text{Å}^3$$
$$r(\mathrm{pH2}) = r(\mathrm{pH7})\frac{D(\mathrm{pH7})}{D(\mathrm{pH2})} = (15.90)\left(\frac{14.25}{12.80}\right) = 17.7\ \text{Å}$$
$$v(\mathrm{pH2}) = 23.2\times10^3\ \text{Å}^3$$
$$v(\mathrm{pH2}) - v(\mathrm{pH7}) = \underline{6.4\times10^3\ \text{Å}^3}$$

c)

$$\frac{s(\text{pH7})}{s(\text{pH2})} = \frac{[D(1-\bar{v}\rho)]\text{pH7}}{[D(1-\bar{v}\rho)]\text{pH2}}$$

$$\bar{v} = \frac{vN_0}{M}$$

$$\bar{v}(\text{pH7}) = \frac{(16.8\times10^3\ \text{Å}^3)(10^{-8}\ \text{cm Å}^{-1})^3(6.023\times10^{23}\ \text{mol}^{-1})}{(14000\ \text{g mol}^{-1})}$$

$$= 0.72\ \text{cm}^3\text{g}^{-1}$$

$$\bar{v}(\text{pH2}) = \left(\frac{23.2}{16.8}\right)(0.72) = 0.99\ \text{cm}^3\text{g}^{-1}$$

$$\frac{s(\text{pH7})}{s(\text{pH2})} = \frac{(14.25)[1-(0.72)(1.04)]}{(12.80)[1-(0.99)(1.04)]} = \frac{(14.25)(0.25)}{(12.80)(-0.030)} = \underline{-9.3}$$

Note that in this solvent lactalbumin floats at pH2 instead of sedimenting; s is negative.

22. a) From Eq. (6.53)

In NaSCN $(d\log c/dx^2) = 0.13\ \text{cm}^{-2}$

In NaCl $(d\log c/dx^2) = 0.40\ \text{cm}^{-2}$

$$M = \frac{(2.303)(2)RT}{\omega^2(1-\bar{v}\rho)}\frac{d\log c}{dx^2}$$

$$= \frac{(4.606)(8.314\times10^7\ \text{erg K}^{-1}\ \text{mol}^{-1})(293\ \text{K})}{(2\pi 21380\ \text{min}^{-1})^2(0.01667\ \text{sec}^{-1}\ \text{min})^2[1-(0.709)(1.20)]}\frac{d\log c}{dx^2}$$

$$M = 1.50\times10^5\frac{d\log c}{dx^2}$$

M (in NaSCN) $= 2.0\times10^4\ \text{g mol}^{-1}$

M (in NaCl) $= 6.0\times10^4\ \text{g mol}^{-1}$

b) The two peptides have molecular weights of 9×10^3 and 12×10^3

c) In NaSCN the protein is a heterodimer of the two polypetide chains (M.W. $\equiv 2\times10^4$). In NaCl three of the heterodimers combine to form a heterohexamer.

23. a) Combining Eq.(6.5) and (6.44):

$$s = \frac{M(1-\bar{v}_\rho)}{N_0 f} = \frac{M(1-\bar{v}\rho}{N_0 6\pi\eta r}$$

$$r = \frac{M(1-\bar{v}\rho)}{N_0 6\pi\eta r}$$

$$= \frac{(310\times10^3\ \text{g mol}^{-1})[1-0.732(1)]}{(6.023\times10^{23}\ \text{mol}^{-1})(6\pi)(1.005\times10^{-2}\ \text{g cm}^{-1}\text{sec}^{-1})(11.7\times10^{-13}\ \text{sec})}$$

$$r = 6.2\times10^{-7}\ \text{cm} = \underline{62\ \text{Å}}$$

b)

$$r(ligated) = \frac{(310\times10^3\ \text{g mol}^{-1})[1-0.732(1)]}{(6.023\times10^{23}\ \text{mol}^{-1})(6\pi)(1.005\times10^{-2}\ \text{g cm}^{-1}\,\text{sec}^{-1})(12.01\times10^{-13}\,\text{sec})}$$

$$r(ligated) = 6.06\times10^{-7}\,cm = 60.6\text{Å}$$

c) The molecule is not spherical and a change in shape occurs.

24. a)

$$M = \frac{RTs}{D(1-\bar{v}\rho)}$$

$$= \frac{(8.314\times10^7)(293)(1025\times10^{-13})}{(3.60\times10^{-8})[1-(0.605)(0.998)]}$$

$$M = \underline{1.76\times10^8\ \text{g mol}^{-1}}$$

b) For no hydration $\bar{v}_2 = v_2$

$$v = \frac{M}{N_0}\bar{v}_2 = \frac{(1.76\times10^8\ \text{g mol}^{-1})(0.605\ \text{cm}^3\,\text{g}^{-1})}{6.023\times10^{23}\ \text{mol}^{-1}}$$

$$v = \underline{1.77\times10^{-16}\ \text{cm}^3}$$

c)

$$f = \frac{kT}{D} = \frac{(1.381\times10^{-16}\ \text{erg mol}^{-1}\,\text{K}^{-1})(293)}{(3.60\times10^{-8}\ \text{cm}^2\,\text{s}^{-1})} = \underline{1.12\times10^{-6}\ \text{g s}^{-1}}$$

d)

$$v = \frac{4}{3}\pi r^3$$

$$r = \frac{f}{6\pi\eta} = \frac{1.12\times10^{-6}}{6\pi(0.01005)} = \underline{5.91\times10^{-6}\ \text{cm}}$$

$$v = \frac{4}{3}\pi r^3 = 8.74\times10^{-16}\ \text{cm}^3$$

e)

$$\frac{V(\text{solvated})}{V(\text{unsolvated})} = \frac{\bar{v}_2 + \delta_1 v_1^{\,0}}{\bar{v}_2} = 4.89$$

$$1 + \delta_1\frac{v_1^{\,0}}{\bar{v}_2} = 4.89$$

$$\delta_1 = 3.89\left(\frac{\bar{v}_2}{v_1^{\,0}}\right) = 3.89\left(\frac{0.605}{0.998}\right) = 2.4\ \text{g H}_2\text{O}(\text{g particle})^{-1}$$

25.

$$\eta = 2.5\bar{v}_2 = \frac{(2.5)(4\pi)(100\times10^{-9}\ \text{cm})(6.023\times10^{23}\ \text{mol}^{-1})}{(1.25\times10^6\ \text{g mol}^{-1})}$$

$$= 2.02\ \text{cm}^3\,\text{g}^{-1}$$

26. a) $$s = \frac{M(1-\bar{v}\rho)}{N_0 f}$$

It is impossible to tell. The density and viscosity of the solvent will decrease. The frictional coefficient of the particle can increase, decrease or remain the same.

b) The frictional coefficient will decrease and thus the sedimentation coefficient will increase.

c) The molecular weight will increase and thus S will increase.

d) $D = kT / f$

When T increases, D will depend on changes in f, so we can not predict change in D . If prolate ellipsoid is cut, D will increase. If molecular weight changes with no change in volume or shape, D will not change.

27. $$M = \frac{RTs}{D(1-\bar{v}\rho)}$$

$$M(\text{prothrombin}) = \frac{(8.314\times10^{7}\ \text{erg mol}^{-1}\ \text{K}^{-1})(293\ \text{K})(4.85\times10^{-13}\ \text{sec})}{(6.24\times10^{-7}\ \text{cm}^2\ \text{sec}^{-1})[1-(0.70)(0.9998)]}$$

$$= 63100\ \text{g mol}^{-1}$$

$$\frac{D_2}{D_1} = \frac{f_1}{f_2} = \frac{r_1}{r_2} = \left(\frac{M_1}{M_2}\right)^{1/3}$$

$$\frac{M_1}{M_2} = \left(\frac{D_2}{D_1}\right)^3 = \left(\frac{6.24}{8.76}\right)^3 = 0.361$$

$$M(\text{thrombin}) = (0.361)(64100) = 22800\ \text{g mol}^{-1}$$

$$M(\text{peptide}) = 63100 - 22800 = 40300$$

28. a) and b) See text

c) Sedimentation and diffusion, or sedimentation equilibrium

d) If the kinetics of the equilibrium is slow compared to electrophoresis rates, a product can be separated from the reactants. The relative amounts of material in the gel bands provide the concentrations necessary to obtain an equilibrium constant.

29. $$s = \frac{M(1-\bar{v}\rho)}{N_0 f};\ D = \frac{kT}{f};\ \mu = \frac{\text{ZeE}}{f}$$

a) As M doubles, s doubles, but D and μ are constant.

b) As f doubles, s, D, and μ are all halved.

c) As Z doubles, μ doubles, but s and D are constant.

d) As f decreases, s, D and μ all increase.

e) As hydration increases, f increases. The values of s, D and μ all decrease.

CHAPTER 7

1. a) $-d[I_2]/dt = k[I_2]^a[\text{ket}]^b[H^+]^c$

Order of reaction with respect to (I_2) is 0.

$$\frac{7\times10^{-5}}{7\times10^{-5}} = \left(\frac{5\times10^{-4}}{3\times10^{-4}}\right)^a$$

$$1 = (1.67)^a \quad \underline{a = 0}$$

Order of reaction with respect to (ket) is 1.

$$\frac{7\times10^{-5}}{1.7\times10^{-4}} = \left(\frac{0.2}{0.5}\right)^b$$

$$0.41 = (0.4)^b \quad \underline{b = 1}$$

Order of reaction with respect to (H^+) is 1.

$$\frac{1.7\times10^{-4}}{5.4\times10^{-4}} = \left(\frac{10^{-2}}{3.2\times10^{-2}}\right)^c$$

$$0.31 = (0.31)^c \quad \underline{c = 1}$$

b) $-d[I_2]/dt = k[\text{ket}][H^+]$

$$k = (7\times10^{-5}/(0.2\times10^{-2}) = 0.035$$

$$k = (1.7\times10^{-4})/(0.5\times10^{-2}) = 0.034$$

$$k = (5.4\times10^{-4})/[(0.5)(3.2\times10^{-2})] = 0.034$$

$$k = \underline{0.034\ \text{M}^{-1}\,\text{s}^{-1}}$$

c) The concentration of ketone will stay essentially constant because $0.5 \gg 10^{-4}$.

$$d[\text{Iket}]/dt = -d[I_2]/dt = (0.034)(0.5)(10^{-1}) \cong \Delta[\text{Iket}]/\Delta t$$

$$\Delta t = \frac{10^{-4}}{(0.034)(0.5)(10^{-1})} = 0.059\ \text{s}$$

Doubling concentration of ket or H^+ will double rate; doubling I_2 will produce no effect.

It is impossible to synthesize 10^{-1} M Iket_1 because the 10^{-3} M I_2 limits the maximum concentration that can be formed.

d) To be consistent with the velocity expression

$$v = k[\text{ket}][H^+]$$

the step involving I_2 should not be rate-determining

$$\text{ketone} + H^+ \rightarrow H^+ \cdot \text{ketone} \qquad \text{(slow)}$$

$$H^+ \cdot \text{ketone} + I_2 \rightarrow \text{iodoketone} + H^+ \qquad \text{(fast)}$$

2. a) Because the half-life decreases twofold as the concentration of each reactant is doubled, the overall kinetic order is 2.

b) $k = \dfrac{1}{(1000\,\text{s})(0.010\,\text{M})} = 0.10\,\text{M}^{-1}\,\text{s}^{-1}$

c) For first expt $1/(0.0025) = 1/(0.005) + (0.10)\,t$

$t = \underline{2000\ \text{s}}$

For second expt $1/(0.0025) = 1/(0.010) + (0.10)\,t$

$t = \underline{3000\ \text{s}}$

d) Possible rate laws are

$v = k[OH^-]^2$

$v = k[OH^-][CH_3COOC_2H_5]$

$v = k[CH_3COOC_2H_5]^2$

or, in general

$v = k[OH^-]^a[CH_3XOOX_2H_5]^b$ where $a + b = 2$

e) Study the half-life for a third experiment where

$[OH^-]_0 = 0.005\,\text{M}$ and $[CH_3COOC_2H_5] = 0.010\,\text{M}$

In fact, any experiment where the two initial concentrations are significantly different from one another would suffice.

3. a) $v = -d[I^-]/dt = k[I^-]^a[OCl^-]^b[OH^-]^C$

$$\frac{1.8\times10^{-4}}{3.6\times10^{4}} = \left(\frac{2\times10^{-3}}{4\times10^{-3}}\right)^a$$

$0.5 = (0.5)^a \qquad \underline{a = 1}$

$$\frac{3.6\times10^{-4}}{7.2\times10^{-4}} = \left(\frac{1.5\times10^{-3}}{3\times10^{-3}}\right)^b$$

$0.5 = (0.5)^b \qquad \underline{b = 1}$

$$\frac{1.8\times10^{-4}}{7.2\times10^{-4}} = \left(\frac{2\times10^{-3}}{4\times10^{-3}}\right)\left(\frac{2.00}{1.00}\right)^c$$

$0.25 = (0.5)(2)^c \qquad \underline{c = -1}$

b) $k = (1.8\times10^{-4}\,\text{M}\,\text{s}^{-1})/[(2\times10^{-3}\,\text{M})(1.5\times10^{-3}\,\text{M})(1.00\,\text{M})^{-1}]$

$= \underline{60\,\text{s}^{-1}}$ First-order overall

c) $v = -d[\mathrm{I^-}]/dt = k_2[\mathrm{I^-}][\mathrm{HOCl}]$

but $[\mathrm{HOCl}] = \mathrm{K_1}[\mathrm{OCl^-}][\mathrm{H_2O}]/[\mathrm{OH^-}]$

Therefore

$v = k_2'[\mathrm{I^-}][\mathrm{OCl^-}]/[\mathrm{OH^-}]$ where $k_2' = k_2[\mathrm{H_2O}]$

Yes, the mechanism is consistent with the rate law.

4. a) $-d[\mathrm{A}]/dt = k_1[\mathrm{A}]$

b) $d[\mathrm{B}]/dt = k_1[\mathrm{A}] - k_2[\mathrm{B}][\mathrm{C}]$

c) $d[\mathrm{D}]/dt = k_2[\mathrm{B}][\mathrm{C}]$

d) $[\mathrm{A}] = [\mathrm{A}]_0\, e^{-k_1 t}$

5. a) $d[\mathrm{C}]/dt = k[\mathrm{A}]^a[\mathrm{B}]^b$

$$\frac{4\times 10^{-3}}{10^{-3}} = \left(\frac{2}{1}\right)^a$$

$4 = 2^a$ $\underline{a = 2}$

b) Order of reaction with respect to [A] is 2

$$\frac{10^{-3}}{10^{-3}} = \left(\frac{2}{1}\right)^b \qquad \underline{b = 0}$$

Order of reaction with respect to [B] is 0

c) $d[C]/dt = k[\mathrm{A}]^2$

d) $k = (10^{-3}\,\mathrm{M\,s^{-1}})/(1\,\mathrm{M})^2 = \underline{10^{-3}\,\mathrm{M^{-1}\,s^{-1}}}$

e) $\mathrm{A} + \mathrm{A} \xrightarrow{k_1} \mathrm{A_2}$ (slow)

$\mathrm{A_2} + \mathrm{B} \xrightarrow{k_2} \mathrm{C} + \mathrm{D} + \mathrm{A}$ (fast)

First reaction is rate determining. Rate of formation of C is essentially equal to rate of formation of $\mathrm{A_2}$.

$d[\mathrm{C}]/dt = d[\mathrm{A_2}]/dt = k_1[\mathrm{A}]^2$

6. $v = -d[\mathrm{S}]/dt = k$

$-d[\mathrm{S}] = k dt$

$\int d[\mathrm{S}] = [\mathrm{S}] - [\mathrm{S}]_0 = -k \int dt = -k(t-0)$

$1/2[\mathrm{S}]_0 - [\mathrm{S}]_0 = -kt_{1/2}$

$t_{1/2} = [\mathrm{S}]_0/2\mathrm{k}$

7. a) $d[\mathrm{D}]/dt = k_3[\mathrm{B}][\mathrm{C}]$

$$d[\mathrm{B}]/dt = k_1[\mathrm{A}] - k_2[\mathrm{B}] - k_3[\mathrm{B}][\mathrm{C}] \cong 0$$
$$k_1[\mathrm{A}] = (k_2 + k_3[\mathrm{C}])[\mathrm{B}]$$
$$[\mathrm{B}] = k_1[\mathrm{A}]/(k_2 + k_3[\mathrm{C}])$$
$$d[\mathrm{D}]/dt = k_3 k_1[\mathrm{A}][\mathrm{C}]/(k_2 + k_3[\mathrm{C}])$$

b) $d[\mathrm{D}]/dt = k[\mathrm{AB}][\mathrm{C}]$

$$K = [\mathrm{AB}]/[\mathrm{A}][\mathrm{B}]$$
$$d[\mathrm{D}]/dt = kK[\mathrm{A}][\mathrm{B}][\mathrm{C}]$$

8. a) Only plot of $1/c$ vs. time is linear.

b) The data best fits second order kinetics.
A least squares fit to the data gives:

$$(1/c) = 4.738\times10^{-2}\,\mathrm{t} + 8.917\times10^{-2}$$
$$(1/c) = kt + 1/c_0$$

Therefore $k = 4.738\times10^{-2}\ \mathrm{mM^{-1}\,s^{-1}}$
Mechanism: $\mathrm{A} + \mathrm{A} \rightarrow \mathrm{B}$

c) A rapid mixing method with absorption detection of concentration would be appropriate.

9. a) $\ln k = \ln A - E_a/RT$

Least squares fit to the data gives:

	$\ln A$	Ea(kJ mol^{-1})
k_1	53.14	106
k_{-1}	95.93	244

$$\Delta H'' = E_a - RT$$
$$\Delta S'' = R\left(\ln\frac{A}{T} - 1 - \ln\frac{R}{\mathrm{Noh}}\right)$$

$T \cong 312$ (average of the 4 temperatures)

	$\Delta H''$(kJ mol^{-1})	$\Delta S''$(J K^{-1} mol^{-1})
k_1	103	188
k_{-1}	241	544

b) $K = k_1/k_{-1}$

$$\ln K = \frac{\Delta S^\circ}{R} - \frac{\Delta H^\circ}{RT}$$

Least squares fit to the data gives:

$$\Delta S^\circ = -356\ \mathrm{J\,K^{-1}\,mol^{-1}}$$
$$\Delta H^\circ = -138\ \mathrm{kJ\,mol^{-1}}$$

The same result is obtained from

$$\Delta H^\circ = \Delta H''(k_1) - \Delta H''(k_{-1})$$

$$\Delta S^\circ = \Delta S''(k_1) - \Delta S''(k_{-1})$$

c) Adding one base pair should not change E_a for forward reaction. It will stabilize the helix and thus increase E_a for backward reaction and make the standard enthalpy more negative.

10. a) $[A] = [A]_0 e^{-kt}$

$$0.10 = e^{-k(1\ \text{hr})}$$

$$k = -\frac{\ln 0.10}{1\ \text{hr}} = 2.30\ \text{hr}^{-1}$$

At 2 hr $[A]/[A]_0 = e^{-k(2\ \text{hr})} = e^{-4.60} = \underline{0.010}$

b) $1/[A] - 1/[A]_0 = kt$

$$1/(0.10[A]_0) - 1/[A]_0 = k(1\,\text{hr})$$

$$k = 9/[A]_0\ \text{hr}^{-1}$$

At 2 hr, $1/[A] = 1/[A]_0 + k(2\,\text{hr})$

$$= 1/[A]_0 + (9/[A]_0)(2) = 19/[A]_0$$

$$[A]/[A]_0 = 1/19 = \underline{0.0526}$$

c) $[A]_0 - [A] = kt$

$$[A]_0 - 0.1[A]_0 = k(1\,\text{hr})$$

$$k = 0.9[A]_0\ \text{hr}^{-1}$$

At 2 hr, $[A] = [A]_0 - k(2\,\text{hr}) = [A]_0 - 1.8[A]_0 = -0.8[A]_0$

The concentration will reach zero before 2 hr elapses.
For all zero-order reactions the rate law must eventually change as concentrations approach zero.

d) $-d[A]/dt = k[A][B]^{1/2}$

but $[A]_0 = [B]_0$ and $[A] = [B]$ at all times

$$-d[A]/dt = k[A]^{3/2}$$

$$\frac{1}{1/2}(1/[A]^{1/2} - 1/[A]_0^{1/2}) = kt \qquad \text{from Eq. (7.25)}$$

$$2[1/(0.1[A]_0)^{1/2} - 1/[A]_0^{1/2}] = 4.325/[A]_0^{1/2} = k(1\ \text{hr})$$

$$k = 4.325[A]_0^{-1/2}\ \text{hr}^{-1}$$

At 2 hr, $2\,((A)^{-1/2} - [A]_0^{-1/2}) = 4.325[A]_0^{-1/2}(2)$

$$[A]^{-1/2} = (4.325 + 1)[A]_0^{-1/2} = 5.325[A]_0^{-1/2}$$

$$([A]/[A]_0)^{1/2} = 1/5.325$$

$$[A]/[A]_0^{1/2} = (1/5.325)^2 = \underline{0.0353}$$

11. a) No change. $[A]'/[A]'_0 = \underline{0.010}$

b) $[A]'_0 = [A]_0/2; k = 9/[A]_0 \text{ hr}^{-1}$

$1/[A]' = 1/[A]'_0 + kt = 1/[A]'_0 + 9/2[A]'_0 = 11/2[A]_0$

$[A]'/[A]'_0 = 1 = \underline{0.182}$

c) $[A]'_0 = [A]_0/2; k = 0.9[A]_0 \text{ hr}^{-1}$

$[A]' = [A]'_0 - kt = [A]'_0 - 1.8[A]'_0 = -0.80[A]'_0$

$[A]'$ is zero before the end of 1 hr.

d) $[A]'_0 = [A]_0/2; k = 4.235[A]_0^{-1/2} \text{ hr}^{-1}$

$$([A]')^{-1/2} = ([A]'_0)^{-1/2} + kt/2 = ([A]'_0)^{-1/2} + \frac{4.325}{2}(2[A]'_0)^{-1/2}$$

$$([A]'/[A]'_0)^{1/2} = 1 + \frac{4.325}{2}(2)^{-1/2}$$

$[A]'/[A]'_0 = 0.156$

12. a) $d[P]/dt = k_3[A_2][C]$

$[A_2] = K_1[A]^2; [C] = K_2[A][B]$

$d[P]/dt = \underline{k_3K_1K_2[A]^3[B]}$

b) (i) $v_0 = k[A]_0^3[B]_0$

$v'_0 = k(2[A]_0)^3(2[B]_0) = 16k[A]_0^3[B]_0 = 16v_0$

(ii) Since $[A]'_0 = [B]'_0$ and $[A]' = [B]'$

therefore $t1/2 \sim 1/[A]_0^3$ Eq. (7.26) with $n = 4$

therefore $t'1/2 \sim 1/(2[A]_0)^3 = 1/8(A)_0^3$

$t'1/2 = t1/2/8$

(iii) $v'_0 = k[A]_0^3(10[B]_0) = 10k[A]_0^3[B] = 10v_0$

13. $(A)/(A)_0 = 2^{-t/t1/2} = 2^{-t/12.5 \text{ yr}}$

$\log 0.2 = (-t/12.5 \text{ yr}) \log 2$

$t = -12.5 \log 0.2/\log 2 = \underline{29 \text{ yr}}$

14. a) $(d[A]/dt)/(d[A]/dt)_0 = k[A]/k[A]_0 = 2^{-t/t1/2}$

$(0.25\times10^5)/(1\times10^5) = 1/4 = 2^{-28d/t1/2} = 2^{-2}$

$t1/2 = 28d/2 = \underline{14 \text{ d}}$

b) $$-(d[A]/dt)_0 = 10^5 \text{ min}^{-1} = k[A]_0 = \frac{0.693}{t_{1/2}}[A]_0$$

$$[A]_0 = \frac{10^5 \text{ min}^{-1}}{0.693}(14d)(1440 \text{ min}/d) = \underline{2.91\times10^9 \text{ atoms}}$$

15. a) $d[\mathrm{B}]/dt = k_1[\mathrm{A}] + k_2[\mathrm{C}] - (k_{-1} + k_{-2} + k_4)[\mathrm{B}]$

b) $d[\mathrm{P}]/dt = k_4[\mathrm{B}]$

Because of the principle of microscopic reversibility, Eq. (7.36)

$[\mathrm{B}] = K_1[\mathrm{A}] \quad d[\mathrm{P}]/dt = k_4K_1[\mathrm{A}]$

c) $[\mathrm{C}] = K_3[\mathrm{A}]$

$$[\mathrm{B}] = k_2[\mathrm{C}] = K_2K_3[\mathrm{A}]$$

$$d[\mathrm{P}]/dt = -d[\mathrm{A}]/dt - d[\mathrm{B}]/dt - d[\mathrm{C}]dt$$

$$= -d[\mathrm{A}]dt - K_1 d[\mathrm{A}]/dt - K_3 d[\mathrm{A}]/dt$$

$$= -(1 + K_1 + K_3) d[\mathrm{A}]/dt = k_4K_1[\mathrm{A}]$$

$$-d[\mathrm{A}]/dt = k_4K_1[\mathrm{A}]/(1 + K_1 + K_3)$$

$$[\mathrm{A}] = [\mathrm{A}]_0^{\mathrm{eq}} e^{-k_4K_1t/(1+K_1+k_3)}$$

$$[\mathrm{A}]_0 = [\mathrm{A}]_0^{\mathrm{eq}} + [\mathrm{B}]_0^{\mathrm{eq}} + [\mathrm{C}]_0^{\mathrm{eq}} = (1 + K_1 + K_3)[\mathrm{A}]_0^{\mathrm{eq}}$$

$$[\mathrm{P}] = [\mathrm{A}]_0 - ([\mathrm{A}] + [\mathrm{B}] + [\mathrm{C}])$$

$$= [\mathrm{A}]_0 - (1 + K_1 + K_3)[\mathrm{A}]$$

$$= [\mathrm{A}]_0 - (1 + K_1 + K_3)\{[\mathrm{A}]_0/(1 + K_1 + K_3)\}e^{-k_4K_1t/(1+K_1+K_3)}$$

$$= [\mathrm{A}]_0(1 - e^{-k_4K_1t/(1+K_1+K_3)})$$

16. a) $\dfrac{d[\mathrm{U}]}{dt} = k_0 - k_1[\mathrm{U}]$

b) $\dfrac{d[\mathrm{U}]}{dt} = \dfrac{k_0}{k_1}(k_1e^{-k_1t}) = k_0e^{-k_1t}$

Substitute $[\mathrm{U}] = \dfrac{k_0}{k_1}\ (1 - e^{-k_1t})$ into a)

$$\frac{d[\mathrm{U}]}{dt} = k_0 - k_1\left(\frac{k_0}{k_1}\right)(1 - e^{-k_1t}) = k_0e^{-k_1t}$$

c) $\dfrac{d[\mathrm{U}]}{dt} = 0$ at maximum, therefore maximum occurs as $t \to \infty$

$$[\mathrm{U}]_{\max} = \frac{k_o}{k_1}$$

$$k_1 = \frac{\ln 2}{(0.500\ \mathrm{hr.})(3600\ \mathrm{s\ hr^{-1}})}$$

$$k_1 = 3.85\times10^{-4}\,\mathrm{s^{-1}}$$

$$[\mathrm{U}]_{\max} = \frac{1.00\times10^{-9}\,\mathrm{M\ s^{-1}}}{3.85\times10^{-4}\,\mathrm{s^{-1}}} = \underline{2.60\times10^{-6}\ \mathrm{M}}$$

This is the limit of [U] as t→Infinity

d) $$[\mathrm{U}]=\frac{k_0}{k_1}(1-e^{-k_1 t})$$

$$e^{-k_1 t}=1-\frac{k_1}{k_0}[\mathrm{U}]=1-\frac{(3.85\times10^{-4})(1.00\times10^{-6})}{(1.00\times10^{-9})}=0.615$$

$$t=\frac{-\ln(0.615)}{k_1}=\frac{-\ln(0.615)}{3.85\times10^{-4}\,\mathrm{s}^{-1}}=\underline{1.26\times10^{3}\,\mathrm{s}}$$

e) $$1-\frac{k_1}{k_0}[\mathrm{U}]=1-\frac{(3.85\times10^{-4})(1.00\times10^{-6})}{0.50\times10^{-9}}=0.23$$

$$t=\frac{-\ln(0.23)}{3.85\times10^{-4}\,\mathrm{s}^{-1}}=3.82\times10^{3}\,\mathrm{s}$$

f) For $[\mathrm{U}]_{max}=1.00\mu\mathrm{M}, k_0=k_1[\mathrm{U}]_{max}=\underline{0.385\mathrm{nM\,s}^{-1}}$

17. a) $$\frac{d[\mathrm{TH}^*\cdot\mathrm{A}]}{dt}=k_{tr}[\mathrm{B}][(\mathrm{TH}^*\cdots\mathrm{A})\mathrm{open}]$$

$$\frac{\mathrm{d}[\mathrm{TH}\ \cdots\mathrm{A})^{\mathrm{open}}]}{dt}=0=k_{op}[(\mathrm{TH}\ \cdots\mathrm{A})_{\mathrm{closed}}]-k_{cl}[(\mathrm{TH}\cdots A)_{\mathrm{open}}]-k_{tr}[(\mathrm{TH}\cdots A)_{\mathrm{open}}][\mathrm{B}]$$

$$[(\mathrm{TH}\cdots\mathrm{A})_{\mathrm{open}}]=\frac{k_{op}[(\mathrm{TH}\cdot\mathrm{A})_{\mathrm{closed}}]}{k_{cl}+k_{tr}[\mathrm{B}]}$$

$$\frac{d[\mathrm{TH}^*\cdot\mathrm{A}]}{dt}=\frac{k_{tr}k_{op}[\mathrm{B}][(\mathrm{TH}\cdot\mathrm{A})_{\mathrm{closed}}]}{k_{cl}+k_{tr}[B]}=k_{ex}[\mathrm{TH}\cdot\mathrm{A})_{\mathrm{closed}}]$$

$$k_{ex}=\frac{k_{tr}k_{op}[\mathrm{B}]}{k_{cl}+k_{tr}[\mathrm{B}]}$$

b) From measurements of the exchange rates vs. concentrations of base pairs, k_{ex} is obtained.

$$\frac{1}{k_{ex}}=\frac{k_{cl}}{k_{op}k_{tr}}\left(\frac{1}{[\mathrm{B}]}\right)+\frac{1}{k_{op}}$$

Plot $1/k_{ex}$ vs. $1/[\mathrm{B}]$; intercept is $1/k_{op}$

c) Least squares fit to data gives intercept of 5.10×10^{-3}s. $\mathrm{k}_{op}=196\,\mathrm{s}^{-1}$

18. A plot of $\log(A)$ vs.t is nearly a straight line

a) First order

$k = \text{Slope} = 0.70/700\,\text{s} = \underline{1\times10^{-3}\,\text{s}^{-1}}$

b) $A + B \xrightarrow{k_1} AB$ (slow)

$AB + P \xrightarrow{k_2} AP + B$ (fast)

Steady state approximation

$$d[AB]/dt = k_1[A][B] - k_2[AB][P] \cong 0$$

$$[AB] = k_1[A][B]/k_2[P]$$

$$d[AP]/dt = k_2[AB][P]$$

$$= k_2(k_1[A][B]/k_2[P])[P]$$

$$= k_1[B][A]$$

$$= k[A]$$

where $k = k_1[B] = \underline{1\times10^{-3}\,\text{s}^{-1}}$

c) $\ln(k_2/k_1) = \ln 2 = -(E_a/R)(1/283 - 1/273)$

$E_a = 5.36\,R = \underline{44.6\,\text{kJ mol}^{-1}}$

19. a) $\Delta H^{\ddagger} = \underline{164\,\text{kJ mol}^{-1}}$

$$\Delta S^{\ddagger} = R(\ln\ Ah/kT - 1)$$

$$= R[\ln\ (1.76\times10^{-11}\times3.2\times10^{16}/383) - 1]$$

$$= (8.314\ \text{J K}^{-1}\ \text{mol}^{-1})(7.29)$$

$$= \underline{60.7\ \text{J K}^{-1}\ \text{mol}^{-1}}$$

b) There are 4 types of covalent bonds present, with bond dissociation energies

$D(C-C = 344\,\text{kJ mol}^{-1}$

$D(C-H) = 415\,\text{kJ mol}^{-1}$

$D(C-O) = 350\,\text{kJ mol}^{-1}$

$D(O-O) = 143\,\text{kJ mol}^{-1}$

The experimental activation enthalpy of $164\,\text{kJ mol}^{-1}$ is sufficient to break only the $O-O$ bond, which is presumed to be the first step in the reaction.

The fact that $\Delta S^{\ddagger}$ is positive and rather large (comparable with vaporizing a mole of liquid) is consistent with a rate-limiting (first) step where two dissociated fragments are formed from a single reactant molecule.

c) $$\ln(k_2/k_1) = \ln 10 = -\frac{164\text{ kJ mol}^{-1}}{R}\left(\frac{1}{T^2} - \frac{1}{383}\right)$$

$$T_2^{-1} = 1/383 - \frac{R\ln 10}{164\text{ kJ mol}^{-1}} = 2.49\times10^{-3}$$

$$T_2 = 401\text{ K} = \underline{128^\circ\text{C}}$$

20. $$\ln\ (k_2/k_1) = \ln 2 = -(E_a/R)(1/308 - 1/298)$$

$$E_a = 6362\,R = \underline{52.9\text{ kJ mol}^{-1}}$$

21. a) $$d[\text{P}]/dt = k[\text{A}][\text{B}]$$

$$= (10^5)(0.1)(0.1)$$

$$= \underline{10^3\text{ M s}^{-1}}$$

b) $$(10^5)(10^{-4})(10^{-6}) = \underline{10^{-5}\text{ M s}^{-1}}$$

c) $$1/[\text{A}] - 1[\text{A}]_0 = kt$$

$$20 - 10 = 10^5 t; \quad t = \underline{10^{-4}\text{ s}}$$

d) $$\ln(k_2/k_1) = \ln 10^3 = -(E_a/R)\ (1/400 - 1/300)$$

$$E_a = 8289\,R = \underline{68.9\text{ kJ mol}^{-1}}$$

$$\Delta H^\ddagger = E_a - RT$$

$$= 68.9 - (8.314)(0.300)$$

$$= \underline{66.4\text{ kJ mol}^{-1}}$$

22. a) $$-d[\text{A}]/dt = k_1[\text{A}][\text{B}]$$

but [B] is constant, so

$$-d[\text{A}]/[\text{A}] = k_1[\text{B}]dt$$

$$\ln([\text{A}]/[\text{A}]_0) = -k_1[\text{B}]t$$

The reaction is first order in [A] and in [B].

b) $-d[\mathrm{A}]/dt = k_2[\mathrm{B}]$, where [B] is constant

$$-d[A] = k_2[\mathrm{B}]dt$$

$$[\mathrm{A}]_0 - [\mathrm{A}] = k_2[\mathrm{B}]t$$

The reaction is zero order in [A] and first order in [B]

c)

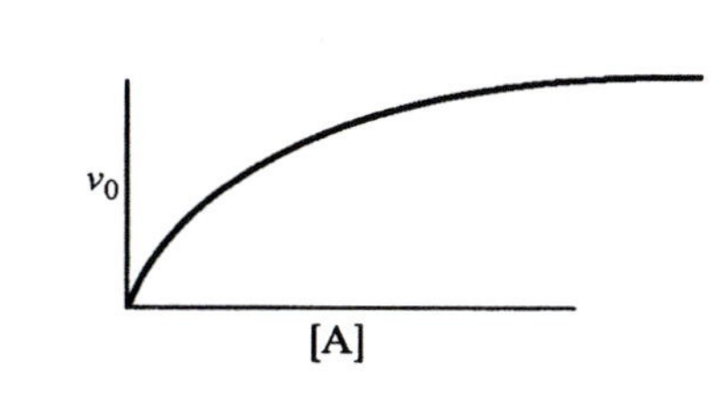

$$v_0 = (-d[\mathrm{A}]/dt)_0$$

$$-d[\mathrm{A}]/dt = k_1[\mathrm{A}][\mathrm{B}]/(1 + k_1[\mathrm{A}]/k_2)$$

For low [A] when $k_1[\mathrm{A}]/k_2 \ll 1, (-d[\mathrm{A}]/dt)_0 = k_1[\mathrm{A}][\mathrm{B}]$

For high [A] when $k_1[\mathrm{A}]/k_2 \gg 1, (-d[\mathrm{A}]/dt)_0 = k_2[\mathrm{B}]$

This behavior is typical of enzyme catalyzed reactions.

23. a) If we express the rate of the reaction as the number of molecules reacting per unit time, then for a unimolecular reaction

$$\mathrm{A} \xrightarrow{k_1} \mathrm{B}$$

$$-dn_A/dt = -(V/N_0)d[\mathrm{A}]/dt = (V/N_0)k_1[\mathrm{A}] = k_1 n_A$$

which is independent of concentration. For a bimolecular reaction

$$2\mathrm{A} \xrightarrow{k_2} \mathrm{B}$$

$$-dn_A/dt = -(V/N_0)d[\mathrm{A}]/dt = (V/N_0)k_2[\mathrm{A}]^2$$

$$= (N_0/V)k_2 n_A^2$$

which still depends on concentration (volume).

b) Based on reasoning analogous to that in part (a), we conclude that it is impossible for the simple mechanism

$$\mathrm{A} \underset{k_2}{\overset{k_1}{\longleftrightarrow}} \mathrm{B}$$

to be first order for the forward reaction and zero order for the reverse reaction. If the system is allowed to come to equilibrium, then the thermodynamic equilibrium constant

$$K_{\mathrm{eq}} = [\mathrm{B}]^{\mathrm{eq}}/[\mathrm{A}]^{\mathrm{eq}}$$

must be independent of dilution. If the kinetics was as proposed, then the relation

$$v_1 = k_1[\mathrm{A}]_{\mathrm{eq}} = k_2 = v_2$$

would violate that requirement.

24. a) Second order

b) $\mathrm{A} + \mathrm{A} \longrightarrow \mathrm{P}$

c) $1/[A]-1/[A]_0 = kt$

$k = [(5\times10^{-3})^{-1} - (10\times10^{-3})^{-1}]/20 \text{ min} = \underline{5 \text{ M}^{-1} \text{ min}^{-1}}$

d) $t_{1/2} = [(1.5\times10^{-3})^{-1} - (3.0\times10^{-3})^{-1}]/5$

$= \underline{67 \text{ min}}$

25. a) First order

b) $A + B \longrightarrow P$ (B in large excess)

c) The reaction is first order in B. It is also first order in A. Because of the large excess of B, it is pseudo zero order in B.

d) $[A] = [A]_0 e^{-k[B]t} = [A]_0 e^{-k't}$

If $[B]_0$ is doubled, then k' doubles and the half-time becomes 50 s.

e) $k' = 0.693/(100 \text{ s})$

$k = k'/[B]_0 = 0.693/[(100 \text{ s})(10 \text{ M})] = \underline{6.93\times10^{-4} \text{ M}^{-1} \text{ s}^{-1}}$

26. $^{14}C/^{14}C_0 = 2^{-t/t_{1/2}} = 0.28$

$$\ln(0.28) = \left(\frac{-t}{12.5 \text{ yr}}\right)\ln 2$$

$t = \underline{22.96 \text{ yr}}$

27. a) The radioactive decay and the excretion of ^{131}I can be considered to be two parallel, first order processes

$$^{131}I \xrightarrow{k_1} {}^{131}Xe + \beta^- \qquad (t_{1/2})_1 = 7.80 \text{ d}$$

$$^{131}I \xrightarrow{k_2} \text{Excretion} \qquad (t_{1/2})_2 = 26 \text{ d}$$

$k_{eff} = k_1 + k_2 = 0.693(1/7.8 + 1/26)d^{-1}$

$(t_{1/2})_{eff} = 0.693/k_{eff}$

$= (1/7.8 + 1/26)^{-1}$

$= \underline{6.0 \text{ d}}$

b) For ^{131}I to be cleared from the animal, sufficient time must elapse from the decay and the excretion to decrease the level of ^{131}I sufficiently. After 2 no $= 60$ d, the attenuation is $2^{60/6} = 2^{10} = 1024$ fold. The residual 0.1% from the previous injection should not interfere with the subsequent experiment.

28. a) $\text{Iso} \xleftrightarrow{k} \text{Allo}; K = 1.38$

For 1 mole of Iso, let $x =$ moles of Allo formed. Then at equilibrium

$x^{eq}/(1 - x^{eq}) = 1.38$

$x^{eq} = (1.38/2.38)$

$= 0.580$

At present time $x/(1 - x) = 0.42$

$x = (0.42/1.42) = 0.296$

$x/x^{eq} = (\text{Allo}/\text{Allo}^{eq}) = \underline{0.510}$

b) $(\text{Allo}/\text{Allo}^{eq}) = 1 - e^{-kt}$

$0.51 = 1 - e^{-k(38600\ \text{yr})}$

$k = 1.848 \times 10^{-5}\ \text{yr}^{-1}$

$t_{1/2} = 0.693/(1.848 \times 10^{-5})$

$= \underline{37{,}500\ \text{yr}}$

c) $\ln(k_2/k_1) = -(E_a/R)(T_2^{-1} - T_1^{-1}) = \ln(t_1/t_2)$

$$\ln\frac{125{,}000}{37{,}500} = -\frac{139{,}700}{8.314}(T_2^{-1} - 1/293)$$

$T_2 = 299\ \text{K} = \underline{26°\text{C}}$

29. a) $\dfrac{d[\text{B}]}{dt} = k[\text{A}] + k_2[\text{AH}^+]$

$= k[\text{A}] + k_2 K[\text{H}^+][\text{A}]$

$\dfrac{-d[\text{A}]}{dt} = \dfrac{d[\text{B}]}{dt}$

$= k[\text{A}]\left(1 + \dfrac{k_2 K}{k}[\text{H}^+][\text{A}]\right)$

$k' = \dfrac{k_2 K}{k}$

b) At pH4 $k(1 + k'[\text{H}^+]) = k(1 + 10^5 \times 10^{-4})$

$= 11\text{k}$

$= 0.693/t_{1/2}$

$= 0.693/(5\ \text{min})$

$k = 0.0126\ \text{min}^{-1} = \underline{0.756\ \text{hr}}$

30. a) $\tau^{-1} = k_1 + k_{-1} = (3 \times 10^{-3}\ \text{s})^{-1}$

$K = [\text{B}]^{eq}/[\text{A}]^{eq} = 10 = k_1/k_{-1}$

therefore $k_1 + k_{-1} = 10k_{-1} + k_1 = 11k_{-1} = (3 \times 10^{-3}\ \text{s})^{-1}$

$k_{-1} = \underline{30.3\ \text{s}^{-1}}; k_1 = 10(30.3\ \text{s}^{-1}) = \underline{303\ \text{s}^{-1}}$

b) $T = \underline{28°\text{C}}$

c) Doubling the concentration of tRNA should have no effect on τ, k_1 or k_{-1}

31. a) Plot τ^{-1} vs $([\bar{\text{I}}^-]+[\bar{\text{I}}_2])$ to obtain

$k_1 = \underline{6.4\times10^9\ \text{M}^{-1}\ \text{s}^{-1}}$ and $k_{-1} = \underline{8.2\times10^6\ \text{s}^{-1}}$

b) $K = k_1/k_{-1} = \underline{780\ \text{M}^{-1}}$

c) From Eq. (7.67)

$$k(\text{diffusion}) = \frac{4\pi(r_{I^-}+r_{I_2})(D_{I^-}+D_{I_2})N_0}{1000}$$

$$= 4\pi(2.16+2.52)\times10^{-8}(2.05+2.25)\times10^{-5}\times6.02\times10^{23}/1000$$

$$= \underline{1.5\times10^{10}\ \text{M}^{-1}\ \text{s}^{-1}}$$

The experimental value is about 40% of the diffusion-limited value. Thus, nearly every other encounter of I_2 with I^- must be effective in producing a reaction.

32. a) $\tau^{-2} = (4k_1[\bar{\text{P}}]+k_{-1})^2 = 16k_1^2[\bar{\text{P}}]^2 + 8k_1k_{-1}[\bar{\text{P}}]+k_{-1}^2$

$[\bar{\text{P}}]^2 = K^{-1}[\bar{\text{P}}_2](k_{-1}/k_1)[\bar{\text{P}}_2]$

$\tau^{-2} = 16k_1^2(k_{-1}/k_1)[\bar{\text{P}}_2] + 8\text{k}_1k_{-1}[\bar{\text{P}}]+k_{-1}^2$

$= 8k_1k_{-1}(2[\bar{\text{P}}_2]+[\bar{\text{P}}])+k_{-1}^2$

$= k_{-1}^2 + 8k_1k_{-1}[\text{P}]_t$

b) From a plot of τ^{-2} vs $[\text{P}]_t$ we obtain

Slope $= 1.36\times10^{16}\ \text{M}^{-1}\ \text{s}^{-2} = 8\,k_1k_{-1}$

Intercept $= 4.0\times10^{12}\ \text{s}^{-2} = k_{-1}^2$

$k_{-1} = \underline{2.0\times10^6\ \text{s}^{-1}}; k_1 = (1.36\times10^{16}\ \text{M}^{-1}\ \text{s}^{-2})/(8\times2.0\times10^6\text{s}^{-1})$

$= \underline{8.5\times10^8\ \text{M}^{-1}\ \text{s}^{-1}}$

c) $\Delta G^\circ = -RT\ln(k_1/k_{-1})$

$= -(8.314)(298)\ln[8.5\times10^8/2.0\times10^6]$

$= -15.0\ \text{kJ mol}^{-1}$

33. a) $\tau^{-1} = k_2 + k_1([NH_3]+[H^+])$

At equilibrium $[\overline{NH_3}] = K[\overline{NH_4}^+]/[\overline{H}^+]$

$$= (5.8\times10^{-10})(0.1)/10^{-6}$$
$$= 5.8\times10^{-5}$$
$$k_2 = Kk_1$$
$$\tau^{-1} = k_1[K + (5.8\times10^{-5} + 10^{-6})]$$
$$= 4.3\times10^{10}\ M^{-1}\ s^{-1}(5.8\times10^{-10} + 5.9\times10^{-5})\ M$$
$$= 2.54\times10^{6}\ s^{-1}$$
$$\tau = \underline{0.39\mu s}$$

b) τ becomes larger as $([\overline{NH_3}]+[\overline{H}^+])$ becomes smaller. However, $[\overline{NH_3}]$ becomes larger as $[\overline{H}^+]$ is decreased, because

$$[\overline{NH_3}] = K([\overline{NH_4}]/[\overline{H}^+]$$

The sum will be maximum when $[\overline{H}^+] = [\overline{NH_3}]$

To show this, write τ (or τ^{-1}) as a function of $[\overline{H}^+]$, take the derivative and set it $= 0$

$$\frac{d\tau^{-1}}{d[H^+]} = \frac{d}{d[H^+]}[k_2 + k_1(K[\overline{NH_4^+}]/[\overline{H}^+] + [\overline{H}^+])]:$$
$$= -k^1K[\overline{NH_4^+}]/[\overline{H}^+]^2 + k_1 = 0$$
$$[\overline{H}^+]^2 = K[\overline{NH_4^+}] = (5.8\times10^{-10})(0.1)$$
$$[\overline{H}^+] = 7.6\times10^{-6} = [\overline{NH_3}]$$
$$pH = \underline{5.1}$$

c) The elementary reaction $H^+ + NH_3 \xrightarrow{k_1} NH_4^+$ dominates in the relaxation process. Therefore, measurement of the temperature dependence of the relaxation time will give the temperature dependence of k_1 and, hence, its activation energy. In aqueous solution, the step is more properly written

$$H_3O^+ + NH_3 \xrightarrow{k_1} H_2O + NH_4^+$$

Thus, it involves the transfer of a hydrogen atom from a water molecule to an ammonia molecule. The process is extremely fast; $k_1 = 4.3\times10^{10}\ M^{-1}\ s^{-1}$ is very close to the diffusion limit. Therefore, the activation energy cannot be large; it is certainly less than $50\ kJ\ mol^{-1}$.

d) According to D – H theory, adding NaCl will change the rate constants only if both species are changed. Adding NaCl will have no effect.

34. a) (I) $1/3(d[O_2]/dt = k[O_3]^2$

(II) $K = k_1/k_{-1} = [O_2][O]/[O_3]$

$$1/2(d[O_2]/dt) = k_2[O_3][O]$$
$$= k_2\, K[O_3]^2/[O_2]$$

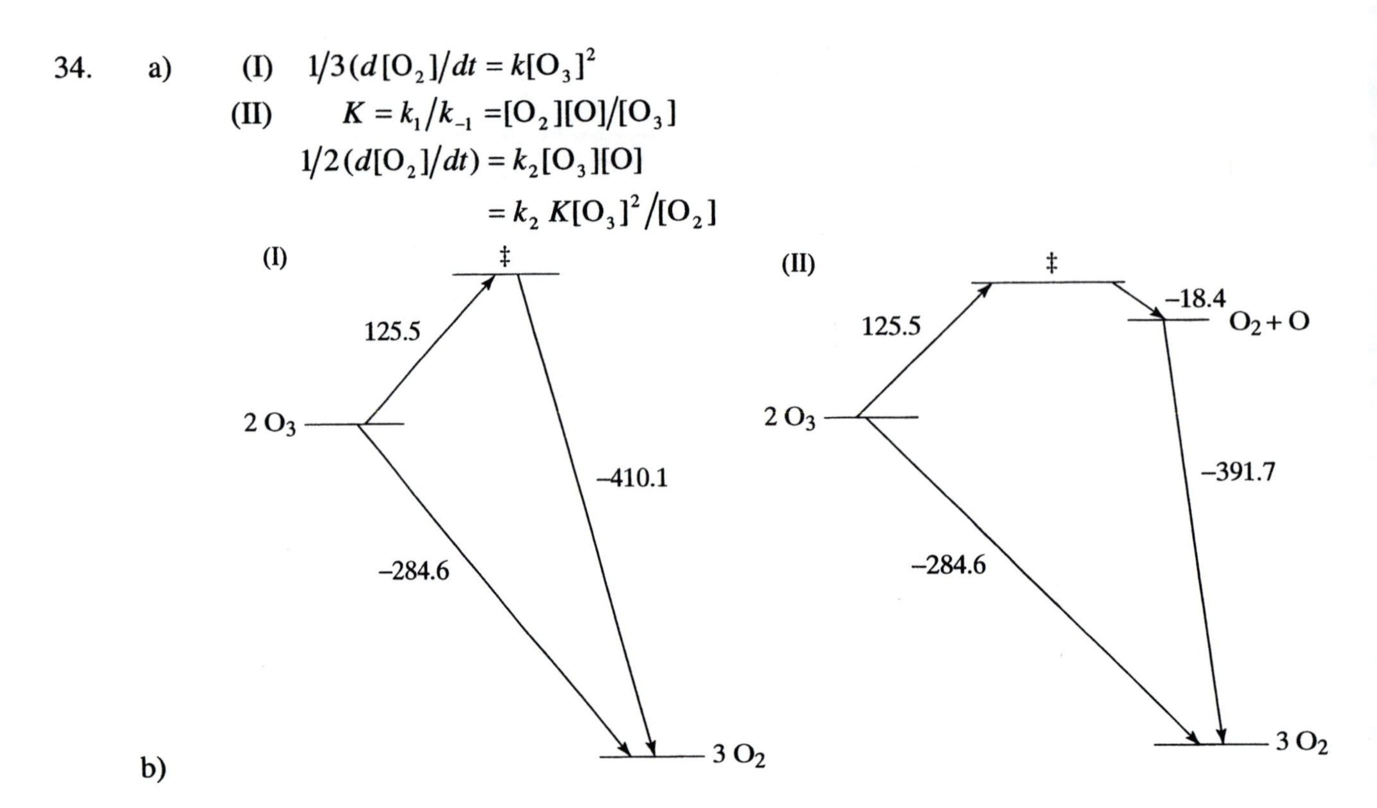

b)

c) The First two steps in Mechanism (II) represent a total of 214.2 kJ/mol, which is greater than the Activation Energy so only Mechanism (I) is possible.

d) Mechanism (II) predicts that the rate of formation of O_2 is inversely proportional to $[O_2]$, while mechanism (I) has a rate independent of $[O_2]$. Simply measure initial rates of formation of O_2 as a function of $[O_2]$ added originally.

35. a) $(S)_1 \underset{k_{-1}}{\overset{k_1\,(\text{slow})}{\longleftrightarrow}} (S)_{m1} \underset{k_2}{\overset{k_2\,(\text{fast})}{\longleftrightarrow}} (S)_{m2} \underset{k_1}{\overset{k_{-1}\,(\text{fast})}{\longleftrightarrow}} (S)_2$

The flux which occurs across the membrane results from a concentration gradient in the direction from left to right

$$(1/A)\, dN_2/dt = k_{-1}[(S)_{m2}] = k_{-1}(k_2/k_2)[(S)_{m1}]$$
$$= (k_1/k_{-1})k_{-1}[(S)_1]$$
$$= k_1[(S)_1]$$

b) $(S)_1 \underset{k_{-1}}{\overset{k_1\,(\text{slow})}{\longleftrightarrow}} (S)_{m1} \underset{k_2}{\overset{k_2\,(\text{fast})}{\longleftrightarrow}} (S)_{m2} \underset{k_1}{\overset{k_{-1}\,(\text{fast})}{\longleftrightarrow}} (S)_2$

$$(1/A)\,dN_2/dt = k_{-1}[(S)_{m2}] = k_{-1}[(S)_{m1}] = 1/2k_{-1}\,[(S)_1]$$

Because the reverse reaction $(S)_{m1} + (S)_1$ occurs at an equal rate to the forward rate $(S)_{m2} + (S)_1$, each will account for half the rate of $(S)_1 + (S)_{m1}$.

c) If the difference between the fast and slow steps is not too great, then we can look at the induction of transport to distinguish between the two mechanisms. In the mechanism of part (a), the concentration of (S) in the membrane will rise slowly and with a constant initial slope; thus the induction of the rate of appearance of $(S)_2$ will be governed by k_1 (slow). In the mechanism of (b), the level of $(S)_m$ will rise quickly and we should not observe an initial induction in the rate of increase of $(S)_2$. The difference is subtle.

d) Electrolyte serves to shield the coulombic interaction between S^+ and m^-. Decreasing the electrolyte concentration then will increase k_1 (attractive interaction) and decrease k_{-1} (separation of opposite charges).

36. a) The coulombic interaction between two like charges is

$$U(r) = \frac{q_1 q_2}{4\pi\varepsilon_0 \in r} \qquad \text{Eq.(9.1)}$$

$$U(r^\ddagger) = \frac{e^2}{4\pi\varepsilon_0 \in r^\ddagger}$$

b) Because the contribution of $U(r^\ddagger)$ is positive (repulsive), it will be decreased in a solvent of high dielectric constant. Thus, the reaction should be faster in a solvent of high dielectric constant, because $\Delta G^{\circ +}$ is decreased.

c) Plot the measured value of $\Delta G^{\circ +}$ for a given reaction in several solvents against $1/\in$. By extrapolating to $E = \infty$, the non-electrostatic contribution is determined. The remainder is electrostatic.

37. a) $K = [\text{Im}][\text{H}^+]/[\text{ImH}^+] = k_{-1}/k_1$
$= (1.5\times10^3\ \text{s}^{-1})/(1.5z10^{10}\ \text{M}^{-1}\ \text{s}^{-1}) = 1.0\times10^{-7}\ \text{M}$

b) $d[\text{ImH}^+]/dt = k_1[\text{Im}][\text{H}^+] - k_{-1}[\text{ImH}^+] + k_2[\text{Im}] - k_{-2}[\text{ImH}^+][\text{OH}^-]$

c) A solution of 0.1 M imidazole at pH7 will contain $0.05\ \text{M ImH}^+$ and 0.05 M Im. These will be the initial concentrations after the shift to pH4. The values of the four terms in part (b) are then:

$7.5\times10^4, 75, 125, 0.125$

The fastest step is rate determining; hence, the direct reaction of Im with H^+ will dominate in the relaxation.

d) $v_0 = (d[\text{ImH}^+]/dt)_0 = 7.5\times10^4\ \text{M s}^{-1}$

e) Both k_1 and k_{-1} should increase with increasing temperature, but k_1 has a value that corresponds to diffusion controlled rates. It should change less with temperature. The greater temperature dependence of k_{-1} is associated with the activation energy required to break the ImH^+ bond upon dissociation.

f) $K = k_{-1}/k_1 \qquad \Delta H \equiv (E_a)_{-1} - (E_a)_1$

We decided in part (e) that $(E_a)_{-1} > (E_a)_1$; therefore ΔH is positive.

38. a) $d[\text{B}]/dt = k_1 - k_2[\text{B}]$

b) Initial rise will be linear in time, followed by leveling off to a stationary state.

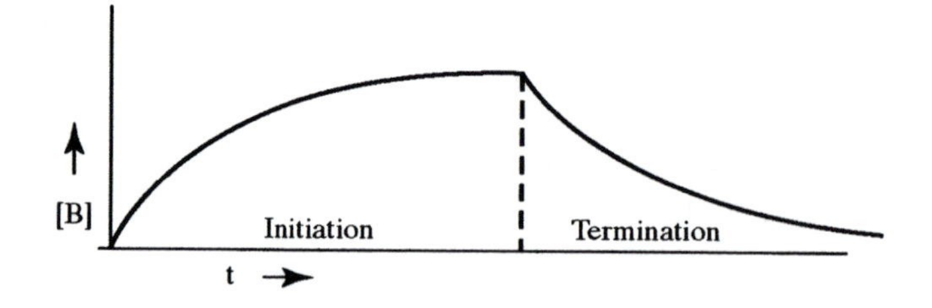

c) Upon termination the concentration of the drug in the blood will decrease exponentially with time.

d) k_2 can be determined from the time required for [B] to drop to half its steady-state value following termination

$k_2 = 0.693/t_{1/2}$

e) k_1 is the initial slope of $[B]$ vs. t. It can be measured at the point of Initiation.

f) Let $x = k_1 - k_2[\mathrm{B}]$, $dx = -k_2 d[\mathrm{B}]$

$d[\mathrm{B}]/dt = -k_2^{-1}\, dx/dt = x$

$\ln(x/x_0) = -k_2 t$

$\ln\{(k_1 - k_2[\mathrm{B}])/(k_1 - k_2[\mathrm{B}]_0)\} = -k_2 t$

But $[\mathrm{B}]_0 = 0$

$\ln(1 - k_2[\mathrm{B}]/\mathrm{k}_1) = -k_2 t$

$[\mathrm{B}] = (k_1/k_2\,(1 - e^{-k_2 t})$

This predicts that $[B]$ approaches a constant value (k_1/k_2) at long times of administration, consistent with the steady state requirement that $v_1 = k_1 = k_2[\mathrm{B}] = v_2$. Upon initiation, the initial slope when t is small can be determined using the series expansion $e^x = a + x + x^2/2! + \cdots$

Thus $[\mathrm{B}]_{t\,\mathrm{small}} = (k_1/k_2)(1 - 1 + k_2 t - \cdots)$

$\equiv k_1 t$

which is consistent with our conclusion of part (e).

g) At steady state $[\mathrm{B}]_{ss} = k_1/k_2$

39. a) Amount of latex produced per acre

$= (10\ \mathrm{kg/tree})(100\ \mathrm{trees/acres})$

$= 10^3\ \mathrm{kg\ acre}^{-1}$

Fuel energy

$= (10^3\ \mathrm{kg\ acre}^{-1}(46{,}000\ \mathrm{kJ\ kg}^{-1}) = 4.6\times10^7\ \mathrm{kJ\ acre}^{-1}$

Solar energy/acre

$= (4\ \mathrm{kJ\ min}^{-1}\ \mathrm{ft}^2)(182{,}500\ \mathrm{min\ yr}^{-1})(43{,}560\ \mathrm{ft}^2\ \mathrm{acre}^{-1})$

$= 3.2\times10^{10}\ \mathrm{kJ\ acre}^{-1}$

Fraction stored

$= (4.6\times10^7)/(3.2\times10^{10}) = 1.4\times10^{-3}$ or <u>0.14%</u>

b) Area needed

$= (14\times10^{16}\ \mathrm{kJ\ yr}^{-1})/(4.6\times10^7\ \mathrm{kJ/acre})(1\ \mathrm{yr})$

$= 3\times10^9\ \mathrm{acre}$

Fraction of US needed

$= (3\times10^9\ \mathrm{acre})/(3\times10^6\ \mathrm{mi}^2\times640\ \mathrm{acre\ mi}^{-2})$

$= 1.56$ or <u>156%</u>

c) Grow the plants in a greenhouse enriched in CO_2. Waste CO_2 produced by combustion or respiration could be used. The temperature and humidity should be controlled to maintain optimum growth. Low grade heat (excess spent cooling water from a nearby power plant) can be used to add the required heat in colder weather.

40. a) $-d[O_3]/dt = k_1[NO][O_3]$

[NO] is in a stationary state, so this is a first-order equation with

$k_{eff} = k_1[NO] = (4\times10^{-15})(10^{10}) = 4\times10^{-5}\,s^{-1}$

$t1/2 = 0.693/k_{eff} = 1.73\times10^4\ s = \underline{4.8\ hr}$

b) Photodecomposition of NO_2 produces NO and O, which have opposite effects on the cycle. NO speeds up step (1) and, hence, the rate of O_3 loss. O reacts with O_2 to form more O_3 by step (B) and, to a lesser extent, with NO_2 by step (2) to deplete NO_2 further. To learn which would predominate, we need more information on the kinetic constants and concentrations.

41. Energy absorbed is $(10\ W)(100\ s) = 10^3\ J$

Number of photons

$= (10^3\ J)(550\times10^{-9}\ m)/[6.62\times10^{-34}\ J\,s)(3\times10^8\ m\,s^{-1})]$

$= 2.8\times10^{21}$

Number of einsteins $= (2.8\times10^{21})/(6.0\times10^{23}) = 4.6\times10^{-3}$

Quantum yield $= (5.75\times10^{-4})/(4.6\times10^{-3}) = \underline{0.125}$

42. a) $$\text{Photons s}^{-1} = \frac{\text{Energy s}^{-1}}{\text{Energy perphoton}} = \frac{\text{Energy s}^{-1}}{hc/\lambda}$$

$= (2\times10^{-16}\ J\,s^{-1})(550\times10^{-9}\ m)/[(6.62\times10^{-34}\ J\,s)(3\times10^8\ m\,s^{-1})]$

$= \underline{554\ s^{-1}}$

b) $\Delta T = (2\times10^{-16}\ J\,s^{-1})(0.24\ cal\ J^{-1})/[(1\ cal\ cm^{-3}\ K^{-1})(10^{-2}\ cm^3)]$

$= 5\times10^{-15}\ deg\ s^{-1}$

c) A temperature rise of 10 K is certainly damaging. Allowing 100 fold safety margin gives $0.1\ K\,s^{-1}$. This is 2×10^{13} times the rise in part (b). Thus, maximum allowable intensity is $(2\times10^{13})\ (2\times10^{-16}\ W) = \underline{4\ mW}$

43. a) $(d[MI]/dt)_0 = k_3(L)$

Now for $BR \xrightarrow{k_2} L$ (fast)

$L \xrightarrow{k_3} MI$ (slow)

$MI \underset{k_{-4}}{\xrightarrow{k_4}} MII$ (very slow)

BR will be converted essentially completely to L before the reaction to form MI gets going.

Thus $[\mathrm{L}]_0 \cong [\mathrm{BR}]_0 \cong [\mathrm{R}]_0$

$(d[\mathrm{MI}]/dt)_0 = -(d[\mathrm{L}]/dt)_0 = k_3[\mathrm{L}]$

$\ln([\mathrm{L}]/[\mathrm{L}]_0) = -k_3 t$ where $k_3 = 10^5\ \mathrm{s}^{-1}$

b) $(d[\mathrm{MI}]/dt)_0 = k_3[\mathrm{L}]_0 \cong (10^5\ \mathrm{s}^{-1})(10^{-3}\ \mathrm{M}) = \underline{10^2\ \mathrm{M\ s}^{-1}}$

Assumes that no *BR* was present prior to illumination and that $k_2 \gg k_3 \gg k_4$

c) $[\mathrm{MII}]/[\mathrm{MI}] \cong k_4/k_{-4} = 1$

Thus $k_{-4} = k_4 = \underline{500\ \mathrm{s}^{-1}}$

d) $k_4/k_{-4} = K_{eq} = 1; \Delta G_4^\circ = 0$

Therefore $\Delta S_{310}^\circ = \Delta H_{310}^\circ / T = (42\ \mathrm{kJ\ mol}^{-1})/(310\ \mathrm{K})$

$= 135.5\ \mathrm{J\ K}^{-1}\ \mathrm{mol}^{-1}$

The large increase in entropy suggests a distinct loss of order in going from MI + MII.

e) $-d[\mathrm{MII}]/dt = (k_{-4} + k_5 + k_6)[\mathrm{MII}] - k_4[\mathrm{MI}]$

$= (k_5 + k_6)[\mathrm{MII}]$ because $k_4[\mathrm{MII}] \cong k_4[\mathrm{MI}]$

44. a) $dN/dt = k_r[A^*]$ where N = no of photons

b) $-d[A^*]/dt = (k_r + k_T)[A^*]$

$\ln([A^*]/[A^*]_0 = -(k_r + k_T)t$

$[A^*]/[A^*]_0 = e^{-(k_r + k_T)t}$

45. Molecules/cm^2 = Photons/cm^2

$$= \frac{(50\times10^{-6}\ \mathrm{J\ s}^{-1}\ \mathrm{cm}^{-2})(300\times10^{-9}\ \mathrm{m})}{(6.62\times10^{-34}\ \mathrm{J\ s})(3\times10^{8}\ \mathrm{m\ s}^{-1})}(1800\ \mathrm{s})$$

$= \underline{1.4\times10^{17}\ \text{molecules cm}^{-2}}$

46. a) $$K = \frac{[\text{complex}]}{[\text{DNA}][\text{psoralen}]}$$

$$250 = \frac{x}{[1.00\times10^{-3} - x][1.00\times10^{-4}]} - x$$

$x = [\text{complex}] = \underline{1.96\times10^{-5}\ \mathrm{M}}$

fraction bound = fraction complex = $\underline{19.6\%}$; fraction free = $\underline{81.4\%}$

b) $$\frac{d(\text{adduct})}{dt} = I_o f(1-10^{-A})(f_{\text{complex}})$$
$$= (1.00\times10^{-11}\ \text{cm}^{-3}\ \text{s}^{-1})(1000\ \text{cm}^3\ \text{L}^{-1})(0.030)(1-10^{-0.3})(0.196)$$
$$= \underline{2.9\times10^{-11}\ \text{M s}^{-1}}$$

c) $f_{\text{complex}} = 0.196$ as before $A = 0.15$ instead of 0.3
$$\frac{d(\text{adduct})}{dt} = \frac{(1-10^{-0.15})}{(1-10^{-0.30})} 2.9\times10^{-11}\ \text{M s}^{-1}$$
$$= \underline{1.7\times10^{-11}\ \text{M s}^{-1}}$$

CHAPTER 8

1. Use Lineweaver-Burk, Eadie-Hofstee or other linear plot to obtain $K_m = 0.42$ M and $V = 0.34$ μmol $CO_2\,min^{-1}$

2. a) $K_m^F = \underline{10\text{ mM}};\ k_2^F = \underline{9\times10^4\ s^{-1}}$

   b) $K_m^R = \underline{12.5\text{ mM}};\ k_2^R = \underline{3\times10^4\ s^{-1}}$

   c) $CO_2 + H_2O + E \underset{k_2^R}{\overset{k_1^F}{\longleftrightarrow}} \text{complex} \underset{k_1^R}{\overset{k_2^F}{\longleftrightarrow}} HCO_3^- + H^+ + E$

   $$K^{eq} = [HCO_3^-][H^+]/[CO_2] = k_1^F\,k_2^F/k_1^R k_2^R$$

   $$K_m^F = (k_2^F + k_2^R)/k_1^F;\quad K_m^R = (k_2^F + k_2^R)/k_1^R$$

   $$K_m^R/K_m^F = k_1^F/k_1^R$$

   $$K^{eq} = K_m^R\,k_2^F/K_m^F k_2^R$$

   The values obtained in parts (a) and (b) refer to pH 7.1
   therefore $K^{eq}/[H^+] = (12.5\text{ mM})(9\times10^{-4}\ s^{-1})/(10\text{ mM})(3\times10^{-4})$
   $= 3.75$
   $K^{eq} = (3.75)(10^{-7.1}) = \underline{3.0\times10^{-7}}$

3. Use Eq. (8.24) or the same method as Prob. 2(c)

   $$K^{eq} = V_F K_m^R/V_L K_m^F = (1.3\times10^3)(1.0\times10^{-5})/[(800)(4.0\times10^{-6})]$$
   $$= \underline{4.1}$$

4. a) $$V = \frac{(2700\text{ units/mg-enzyme})(10\ \mu\text{mol}/15\text{ min/unit})}{[(15\text{ min})(60\text{ s min}^{-1})]}$$

   $$= \underline{3.00\times10^{-5}\ \mu\text{mol } PP_i/(\text{s mg-enzyme})}$$

   b) 1 mg enzyme contains $(10^{-3}\text{ g})(120{,}000\text{ g/mol})^{-1}(6\text{ sites/molecule})$

   $[E]_0 = \underline{5.0\times10^{-8}\text{ mol of sites}}$

   c) $k_2 = V/[E]_0 = (3.00\times10^{-5}\ s^{-1})/5.0\times10^{-8}) = \underline{600\ s^{-1}}$

   d) K_m corresponds to the concentration of substrate where $v_0 = V/2$. Thus, $K_m = \underline{5\times10^{-6}\text{ M}}$.

5. a) $$d[P]/dt = k_2[E]_o/(1 + K_m/[S])$$
   $$= (100\,s^{-1})(10^{-5}M)[1 + (10^{-4})/0.1] = \underline{10^{-3}\ M\,s^{-1}}$$

   b) $$\ln(k_2'/k_2) = -(E_a/R)(1/T' - 1/T)$$
   $$E_a = (8.314)[\ln(200/100)]/(1/280 - 1/300)$$
   $$= \underline{24.2\text{ kJ mol}^{-1}}$$

   c) $K_1^{eq} = k_1/k_{-1} \cong 1/K_m = k_1/(k_2 + k_{-1})$ because $k_2 << k_{-1}$
   $= \underline{10^4\ M^{-1}}$

d) $\ln(K_m/K_m') = \ln(K'/K) = -(\Delta H^\circ/R)(1/T' - 1/T)$

$$\Delta H^\circ = (8.314)[\ln(1.0\times10^{-4}/1.5\times10^{-4})]/(1/280 - 1/300)$$
$$= \underline{-14.16 \text{ kJ/mol}}$$

6. Typically a catalyst functions by decreasing the activation energy for a reaction. When this occurs the reaction becomes less sensitive to temperature.

7. a) $\ln 2.2 = -(E_a/R)(1/298 - 1/283)$

$$E_a = \underline{36.9 \text{ kJ mol}^{-1}}$$

b) $k_2 = (kT/h)\, e^{\Delta S^\ddagger/R} e^{-\Delta H^\ddagger/RT}$

$$\Delta S^\ddagger = R\ln(k_2 h/kT) + \Delta H^\ddagger/T$$
$$= (8.314)\ln\left[\frac{(560 \text{ s}^{-1})(6.62\times10^{-34} \text{ J s})}{(1.38\times10^{-23} \text{ J deg}^{-1})(298)}\right] + \frac{(36{,}900)}{298}$$
$$= \underline{-68.5 \text{ J K}^{-1} \text{ mol}^{-1}}$$

Thus, the transition state for this complex is more ordered and has a higher energy than do the reactants. Presumably this has to do with the energy involved in breaking H-bonds between urea and H_2O and the loss in entropy in binding both urea and H_2O to the enzyme.

c) $\ln(k_{\text{uncat}}/k_{\text{cat}}) = \ln(10^{14}) = -(E_a/R)(1/T_2 - 1/298)$

$$1/T_2 = 1/298 - R\ln 10^{14}/105{,}000$$
$$T_2 = 1245 \text{ K} = \underline{972^\circ\text{C}}$$

8. $d[\text{P}]/dt = k_2[\text{ES}]$

Steady state $d[\text{ES}]/dt = k_1[\text{E}][\text{S}] - (k_1 + k_2)[\text{ES}] = 0$

$$[\text{E}]_0 = [\text{E}] + [\text{ES}] + [\text{EI}]$$
$$K_I = [\text{E}][\text{I}]/[\text{EI}] = k_{-3}/k_3$$
$$[\text{E}]_0 = [\text{E}](1 + [\text{I}]/K_1) + [\text{ES}]$$

Thus $[\text{E}] = (k_{-1} + k_2)[\text{ES}]/k_1[\text{S}] = k_m[\text{ES}]/[\text{S}]$

$$= ([\text{E}]_0 - [\text{ES}])/(1 + [\text{I}]/K_I)$$
$$[\text{ES}] = [\text{E}]_0/\{1 + (1 + [\text{I}]/K_I)K_m/[\text{S}]\}$$
$$\frac{d[\text{P}]}{dt} = \frac{k_2[\text{E}]_0}{1 + (1 + [\text{I}]/K_I)K_m/[\text{S}]}$$

9. a) k_1 characterizes $\text{E} + \text{F} + \text{EF}$

k_2 characterizes $\text{EF} + \text{EM}$

k_{-3} characterizes $\text{E} + \text{M} + \text{EM}$

b) $K = k_1/k_{-1} = [\text{EF}]/[\text{E}][\text{F}]$

The temperature dependence of k_1 and k_{-1} will give ΔH° for $\text{E} + \text{F} + \text{EF}$. Similarly, T-dependence of k_3 and k_{-3} will give ΔH° for $\text{E} + \text{M} + \text{EM}$.

c) From Eq. (8.24) $K = V_F K_m^M / V_M K_m^F$. Therefore, the T dependence of the maximum velocities and Michaelis constants for the forward and reverse reactions can be measured to give ΔH°.

10. Enzyme concentration $= \dfrac{(0.04\ \mu\text{g ml}^{-1})(1000\ \text{ml l}^{-1})}{(30{,}000\ \mu\text{g}/\mu\text{mol})}$

$$= 1.3\ \text{nM} = 1.3\times10^{-9}\ \text{M}$$

a) $[\text{ES}]/[\text{E}]_0 = \{1 + (k_{-1} + k_2)/k_1[\text{S}]\}^{-1} = (1 + K_m/[\text{S}])^{-1}$

$$= \{1 + (5\times10^{-5}\,\text{M})/(5\times10^{-6}\ \text{mol}/10^{-3}\ \text{L})\}^{-1}$$

$$= \underline{0.99}$$

b) By the time [S] has decreased from 5×10^{-3} M to 2.5×10^{-3} M, the velocity will still be 0.98 V; therefore, we can assume that the kinetics is approximately zero order.

$$v_0 \cong V = \frac{\Delta[\text{S}]}{\Delta T} = k_2[\text{E}]_0$$

$$\Delta T = (2.5\times10^{-3}\ \text{M})/(2000\ \text{s}^{-1})(1.3\times10^{-9}\ \text{M})$$

$$= 960\ \text{s} = \underline{16\ \text{min}}$$

c) $v_0 = 1/2\ \text{V}$ when $[\text{S}] = K_m = \underline{5\times10^{-5}}\ \text{M}$

d) Yes. In part (b) the presence of 50% more enzyme will result in the time of reaction being only 2/3 as long, 10.7 min.
For part (c), the value is unaffected because K_m is unchanged.

e) Same affinity implies that $K_I = [\text{E}][\text{I}]/[\text{EI}]$

$$= [\text{E}][\text{S}]/[\text{ES}] = K_m$$

$$v_0 = V/(1 + K_m'/[\text{S}]) = V/5(1 + K_m/[\text{S}])$$

$$K_m' = [\text{S}](5 + 5K_m/[\text{S}] - 1) = 4[\text{S}] + 5K_m = K_m[1 + [\text{I}]/K_m)$$

$$4[\text{S}] + 4K_m = [\text{I}] = 4(K_m + [\text{S}])$$

When $[\text{S}] << K_m$ then $[\text{I}] = 4K_m = 4(5\times10^{-5}\ \text{M}) = \underline{2\times10^{-4}\ \text{M}}$ is required.

11. a) Lineweaver-Burk plot gives $K_m = \underline{44\ \text{mM}}$

b) Lineweaver-Burk plot in the presence of urea gives same K_m value, but different V. Therefore inhibitor is non-competitive.

12.

$$E + S \;\frac{}{+17.6}\; ES \;\frac{}{+25.5}\; (EZ)^{\neq} \;\frac{}{-63.0}\; EP \;\frac{}{+5.0}\; E + P$$

(overall, E + S to E + P: −15.1)

13. a)

$$\begin{array}{ccc} E + S & \underset{}{\overset{K_1}{\rightleftharpoons}} & ES \\ K_2 \updownarrow H^+ & & K_4 \updownarrow H^+ \\ H^+E + S & \underset{}{\overset{K_3}{\rightleftharpoons}} & H^+ES \xrightarrow{k} P + H^+E \end{array}$$

$K_1 = [ES]/[E][S]$; $K_2 = [H^+E]/[E][H^+]$

$K_3 = [H^+ES]/[H^+E][S]$; $K_4 = [H^+ES]/[ES][H^+]$

$[ES] = K_1[E][S] = [H^+ES]/K_4[H^+]$

$K_1K_4 = [H^+ES]/([E][S][H^+])$

$[H^+E] = K_2[E][H^+] = [H^+ES]/K_3[S]$

$K_2K_3 = [H^+ES]/([E][S][H^+])$

$= K_1K_4$

b) $v_0 = d[P]/dt = k[H^+ES]$

$[H^+ES] = K_1K_4[E][S][H^+]$

$[E]_0 = [E] + [ES] + [H^+E] + [H^+ES]$

$= [H^+ES]/K_1K_4[H^+][S] + [H^+]/K_4[H^+]$

$+ [H^+ES]/K_3[S] + [H^+ES]$

$[H^+ES] = [E]_0/\{1/K_1K_4[H^+][S] + 1/K_4[H^+] + 1/K_3[S] + 1\}$

$v_0 = k[E]_0/\{1 + 1/K_4[H^+] + (1/K_1K_4[H^+] + 1/K_3)[S]\}$

$= v'/(1 + K'_m/[S])$

where $v' = k[E]_0/(1 + 1/K_4[H^+])$

and $K'_m = (1/K_1K_4[H^+] + 1/K_3)/(1 + 1/K_4[H^+])$

c) At high substrate concentrations $[S] >> K'_m$ and

$v_0 = V' = kK_4[E]_0[H^+]/(K_4[H^+] + 1)$

$1/V' = (1/k[E]_0)(1 + 1/K_4[H^+])$

Thus, a plot of $1/V'$ vs $1/[H^+]$ should give a straight line.

14. $K_m = \underline{9.1\ \mu\text{M}}$; $V = \underline{80\ \text{nM s}^{-1}}$

Upward curvature of $1/v_0$ vs 1/[S] plot reflects a lower velocity than expected at high substrate concentrations, relative to that expected from the Michaelis-Menten mechanism. This appears to be an example of substrate inhibition at high substrate concentrations. On this basis K_m and V are best determined emphasizing the data at low $[\text{S}]$; i.e., large 1/[S].

15. b) $1/V$ and $1/K_m$ are apparently both zero.

c) $k_2 >> (k_1 + k_{-1})$; ES dissociates to E and P as soon as it is formed.

d) First order, because v_0 is directly proportional to $[\text{O}_2^-]$

e) $v_0 = k[\text{E}]_0[\text{O}_2^-]$

$k[\text{E}]_0 = 5\times10^2\ \text{s}^{-1}$ from slope of v_0 vs $[\text{O}_2^-]$ plot

$$k = (5\times10^2\ \text{s}^{-1})/(4\times10^{-7}\ \text{M}) = \underline{1.25\times10^9\ \text{M}^{-1}\ \text{s}^{-1}}$$

which is a value close to the diffusion limit

f)
$$v_0 = k_1[\text{E}][\text{O}_2^-] + k_2[\text{E}^-][\text{O}_2^-]$$
$$= (k_1[\text{E}] + 2k_1[\text{E}^-])[\text{O}_2^-] = k_1([\text{E}] + 2[\text{E}^-])[\text{O}_2^-]$$

$([\text{E}] + 2[\text{E}^-]) = \text{constant} \times [\text{E}]_0$

because the two terms in the velocity law must be identical.

g) $k_1[\text{E}][\text{O}_2^-] = 2k_1[\text{E}^-][\text{O}_2^-]$; $[\text{E}]/[\text{E}^-] = 2$

h) $k = k_1 + k_2 = k_1 + 2k_1 = 3k_1$;

$k_1 = \underline{4.2\times10^8\ \text{M}^{-1}\ \text{s}^{-1}}$, $k_2 = \underline{8.3\times10^8\ \text{M}^{-1}\ \text{s}^{-1}}$

16. a) $K_1 = [\overline{\text{C}}]/([\overline{\text{A}}][\overline{\text{B}}])$

$$= (1\times10^{-11}\ \text{M})/(6\times10^{-10}\ \text{M})^2 = \underline{2.8\times10^7\ \text{M}^{-1}}$$

$K_2 = [\text{D}]/[\text{C}] = (0.39\ \text{nM})/(0.01\ \text{nM}) = 39$

b)
$$d[C]/dt = k_1[\text{A}][\text{B}] - k_{-1}[\text{C}] + k_{-2}[\text{D}] - k_2[\text{C}]$$
$$= k_1[\text{A}][\text{B}] - k_{-2}[\text{D}] - (k_{-1} + k_2)[\text{C}]$$

$d[\text{D}]/dt = k_2[\text{C}] - k_{-2}[\text{D}]$

c) The first reaction comes quickly to equilibrium with

$K_1 = 2.8\times10^7\ \text{M}^{-1} = x_1/(10^{-9} - x_1)^2$

(1) Assuming that $x_1 << 10^{-9}$

$x_1 = (2.8\times10^7)(10^{-9})^2 = 2.8\times10^{-11}\ \text{M}$

(2) Assuming that $10^{-9} - x = 0.972\times10^{-9}$

$x_1 = (2.8\times10^7)(0.972\times10^{-9})^2 = 2.6\times10^{-11}\ \text{M}$

Thus [A] and [B] will decrease rapidly from 1 nM to 0.974 nM and then slowly to 0.60 nM.

[C] will rise rapidly to 0.026 nM and then slowly decrease to 0.01 nM; it will exhibit a maximum.

[D] will slowly increase from zero to 0.40 nM.

d) $k_{-2} = k_2/K_2 = (0.05\,\text{s}^{-1})/(39) = \underline{1.28\times10^{-3}\ \text{s}^{-1}}$

17. b) $V = \underline{1.23\times10^{-6}\ \text{M s}^{-1}}$; $K_m = \underline{4.85\times10^{-4}\ \text{M}}$

c) In the absence of inhibitor

$$
\begin{aligned}
v_0 &= V/(1+K_m/[\text{S}]) \\
&= V/1+4.85\times10^{-4}/1.0\times10^{-3}) \\
&= V/1.485
\end{aligned}
$$

In the presence of inhibitor

$$v_0 = V/(1+K'_m/[\text{S}]) \text{ where } K'_m = K(1+[\text{I}]/K_1)$$
$$V/2\times1.485 = V/(1+K'_m/10^{-3})$$

$$K'_m = 10^{-3}(2.97-1) = 1.97\times10^{-3}\ \text{M} = (4.85\times10^{-4}\ \text{M})\left(1+\frac{30\times10^{-3}}{K_I}\right)$$

$$K_I = \frac{30\times10^{-3}\ \text{M}}{4.06-1} = 9.8\times10^{-3}\ \text{M}$$

Lineweaver-Burk plot with intercept $1/V = 0.813\times10^6\ \text{M}^{-1}\text{s}$ and slope $K'_m/V = 1602\ \text{s}$

18. a) Curve rises exponentially ($t1/2 = 4$ min) to a maximum concentration of $1\ \text{mol}/40\ l = 0.025\ \text{M}$ and then decreases linearly to zero during next 6 hr.

b) Maximum concentration = $\underline{0.025\ \text{M}}$.

Recuperation time = $\underline{6\ \text{hr}}$.

c) Maximum concentration = $\underline{0.050\ \text{M}}$

Value remains above 0.025 M for 6 hr

Hangover period is $\underline{12\ \text{hr}}$.

d) $\underline{12\ \text{hr}}$

e) Ten ml hr^{-1} is being removed, therefore 10 ml hr^{-1} could be consumed.

f) This tactic has virtually no effect. During 1 hr less than 10% of the 120 ml is metabolized. The intoxication period is actually somewhat longer than if the 120 ml had been consumed initially.

19. a)

$$\text{NAD}^+ + \text{LADH} \underset{k_{-1}}{\overset{k_1}{\longleftrightarrow}} \text{NAD}^+\text{LADH}$$

$$\text{NAD}^+\text{LADH} + \text{C}_2\text{H}_5\text{OH} \underset{k_{-2}}{\overset{k_2}{\longleftrightarrow}} \text{NAD}^+\text{LADH C}_2\text{H}_5\text{OH}$$

$$\text{NAD}^+\text{LADH C}_2\text{H}_5\text{OH} \underset{k_{-3}}{\overset{k_3}{\longleftrightarrow}} \text{NADH LADH} + \text{CH}_3\text{CHO} + \text{H}^+$$

$$\text{NADH LADH} \underset{k_{-4}}{\overset{k_4}{\longleftrightarrow}} \text{NADH} + \text{LADH}$$

b) Either NAD^+ is limiting or the LADH NAD^+ complex is saturated with respect to C_2H_5OH

c) $$\text{LADH Present} = \frac{(4\times10^{-3}\ \text{M hr}^{-1})(40\ 1)(10^6\ \mu\text{mol/mol})}{(3600\ \text{s hr}^{-1})(3.1\ \text{s}^{-1})}$$

$$= \underline{14\ \mu\text{mol}}$$

d) $\tau^{-1} = k_1([\text{LADH}] + [\text{NAD}^+]) + k_{-1}$

Because $[\text{LADH}] << [\text{NAD}^+]$

$$(1.65\times10^{-3}\ \text{s})^{-1} = k_1(1\times10^{-3}\ \text{M}) + k_{-1}$$

$$(7.9\times10^{-3}\ \text{s})^{-1} = k_1(1\times10^{-4}\ \text{M}) + k_{-1}$$

$$k_1 = \underline{5.3\times10^5\ \text{M}^{-1}\,\text{s}^{-1}}$$

$$k_{-1} = \underline{75\ \text{s}^{-1}}$$

e) By competitive inhibition ethanol will prevent methanol or ethylene glycol from being oxidized so rapidly. In time other elimination processes can then remove the toxic substances from the system.

20. b) $K_m(C_2H_5OH) = \underline{0.44\ \text{mM}}$; $K_m(NAD^+) = \underline{18\ \mu\text{M}}$

c) Competitive

d) Non-competitive

e) $NAD^+LADH + \text{pyrazole} \longleftrightarrow NAD^+\,LADH$ pyrazole competes with step 2 of mechanism in Prob 19(A).

f) Pyrazole should slow down ethanol oxidation and, hence, keep the $[NAD^+]/[NADH]$ ratio from falling too low.

21. a) $$v_0 = \frac{K_{cat}[E]_0}{1 + K_M/[S]} = \frac{(150\ \text{s}^{-1})(0.01\ \mu\text{M})}{1 + (2.00\ \mu\text{M})/(5.00\ \mu\text{M})}$$

$$v_0 = \underline{1.07\ \mu\text{M s}^{-1}}$$

b) $$\frac{v_0'}{v_0} = \frac{1 + K_M/[S]}{1 + K_M'/[S]} = \frac{1}{2}$$

$$K_M' = [S] + 2K_M = 5.00\ \mu\text{M} + 2(2.00\ \mu\text{M})$$

$$K_M' = 9.00\ \mu\text{M}$$

$$K_M' = K_M(1 + [I]/K_I)$$

$$K_I = \frac{[I]}{(K_M'/K_M) - 1}$$

$$K_I = \frac{[I]}{(K_M'/K_M) - 1} = \frac{5.00\ \text{mM}}{(9.00\ \mu\text{M}/2.00\ \mu\text{M}) - 1}$$

$$K_I = \underline{1.43\ \text{mM}}$$

c) Use Eq. (8.25)

$$\frac{v_B}{v_A} = \frac{(K_{\text{cat}}/K_M)_B}{(K_{\text{cat}}/K_M)_A} = \frac{(100/10.00)}{(150/2.00)} = \underline{0.133}$$

22. a) Lineweaver-Burk plots give:

$$K_M(20^\circ\text{C}) = 0.468\ \mu\text{M} \quad V_{\max}(20^\circ\text{C}) = 1.20\ \mu\text{M s}^{-1}$$

$$K_M(40^\circ\text{C}) = 0.469\ \mu\text{M} \quad V_{\max}(40^\circ\text{C}) = 5.80\ \mu\text{M s}^{-1}$$

b) $K_{\text{cat}}(20^\circ\text{C}) = V_{\max}/[\text{E}]_0 = 0.12\ \text{s}^{-1}$

$K_{\text{cat}}(40^\circ\text{C}) = 0.58\ \text{s}^{-1}$

c) $$E_a = -R\left(\ln\frac{k_2}{k_1}\right)\left(\frac{1}{T_2} - \frac{1}{T_1}\right)^{-1}$$

$E_a = \underline{60.1\ \text{kJ mol}^{-1}}$

d) The enzyme-catalyzed reaction should have a lower activation energy; that is how enzymes catalyze reactions.

23. $$V_1 = \frac{V}{1 + [\text{I}]/K_1}$$

$$[\text{I}] = K_1[(V/V_1) - 1] = (2.9\times10^{-4})(10-1)$$

$$[\text{I}] = 2.6\times10^{-3}\ \text{M}$$

24. a) $$\frac{d[\text{GES}]}{dt} = k_1[\text{GE}][\text{S}] - (k_c + k_{-1})[\text{GES}]$$

$$\frac{d[\text{GES}]}{dt} = \frac{k_1[\text{E}][\text{G}][\text{S}]}{K_G} - (k_c + k_{-1})[\text{GES}]$$

but $[\text{E}_0] = [\text{E}] + [\text{ES}] + [\text{GE}] + [\text{GES}]$

$$\therefore [\text{E}] = \frac{K_G[\text{E}_0]}{K_G + [\text{G}]} - \frac{K_G}{[\text{G}]}[\text{GES}]$$

and $$\frac{d[\text{GES}]}{dt} = \frac{k_1[\text{E}_0][\text{G}][\text{S}]}{K_G + [\text{G}]} - (k_c + k_{-1} + k_1[\text{S}])[\text{GES}] = 0$$

$$[\text{GES}] = \frac{[\text{E}_0]}{\left(1 + \frac{K_G}{[\text{G}]}\right)\left(1 + \frac{K_M}{[\text{S}]}\right)}$$

$$K_M = \frac{k_c + k_{-1}}{k_1}$$

b) $K_M = \frac{k_c + k_{-1}}{k_1} = \frac{350\ \text{min}^{-1} + 0.2\ \text{min}^{-1}}{9\times10^7\ \text{M}^{-1}\ \text{min}^{-1}} = 3.9\times10^{-6}\ \text{M}$

$K_M = 3.9\,\mu\text{M}$

$$\frac{d[\text{P}]}{dt} = \frac{(0.06\ \text{min}^{-1})(100\,\mu\text{M})}{\left(1+\frac{0.5\ \text{mM}}{[\text{G}]}\right)\left(1+\frac{3.9\,\mu\text{M}}{[\text{S}]}\right)}$$

The rate follows Michaelis-Menten kinetics with $V_{\max}$ dependent on [G].

$$V_{\max} = \frac{k_P[\text{E}_0]}{1+\frac{K_G}{[\text{G}]}}$$

c) Do not add cofactor G so that reaction of $\text{E} + \text{S} \rightarrow \text{ES}$ can be studied separately from the kinetics of product formation. A temperature-jump experiment is appropriate.

25. a) $k = \underline{2\times10^5\ \text{M}^{-1}\ \text{s}^{-1}}$

b) The light pulse must be sufficiently short that it does not provide a limitation to the measurement of the kinetic relaxation in the millisecond regime.

26. a) $v_0 = V/(1 + K_m/[\text{S}])$

Because $K_m >> [\text{S}]$ during the entire course of the reaction

$v = V[\text{S}]/K_m$ (first order) and $\ln([\text{S}]/[\text{S}]_0) = -(V/K_m)t$

$t = -(10^{-3}\ \text{M})(\ln 0.1)/(1\,\text{M s}^{-1}) = 2.3\times10^{-3}\ \text{s} = \underline{2.3\ \text{ms}}$

b) When $K_m << [\text{S}]$, $v_0 = V$

$\Delta t = \Delta S / V = 0.9\ \text{M}/1\ \text{M s}^{-1} = \underline{0.9\ \text{s}}$

c) In the presence of $10^{-4}\ \text{M}\,\text{Mg}^{2+}$,

$K'_m = K_m(1 + [\text{I}]/K_1) = 10^{-3}\ \text{M}[1 + (10^{-4}\ \text{M})/(10^{-4}\ \text{M})]$

$= \underline{2\times10^{-3}\text{M}}$

Conditions of part (a)

$t = -(2\times10^{-3}\ \text{M})(\ln 0.1)/(1\ \text{M s}^{-1}) = \underline{4.6\ \text{ms}}$

Conditions of part (b)

$\Delta t = \underline{0.9s}$ (no effect of inhibitor at high substrate concentration)

27. a) Because $v_0 = V/2$ when $[\text{S}] = 3\times10^{-3}\ \text{M}$

$K_m = \underline{3\times10^{-3}\ \text{M}}$

b) Because $v_0 = V/2$ when $[\text{S}] = 6\times10^{-3}\ \text{M}$

$K'_m = 6\times10^{-3}\ \text{M} = K_m(1 + [\text{I}]/K_I)$

$1 + [\text{I}]/K_I = (6\times10^{-3})/(3\times10^{-3}) = 2$

$K_I = [\text{I}] = \underline{1.0\times10^{-4}\ \text{M}}$

c)

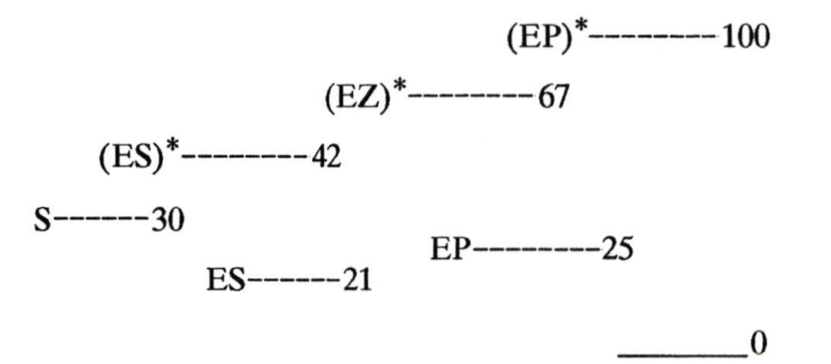

d) $k_3; \Delta H^{\ddagger}_{k_{-1}} = 21$ kJ

e) The step from $EP + P$ is probably rate determining because it has a significantly larger activation energy than any other step.

28. a) The inhibition appears to be competitive

$V = 1.33 \times 10^{-6}$ M min^{-1}

$K_m = 7.9$ mM; $K'_m = 88$ mM

b) $1 + [\text{I}]/K_I = K'_m/K_m = 88/7.9 = 11.1$

$K_I = 0.2 \times 10^{-4}/10.1 = 0.2 \times 10^{-5}$ M

Binding constant $= 1/K_I = \underline{5 \times 10^{-5}\ \text{M}^{-1}}$

29. a) For the three curves shown

$K'_m = K_m(1 + [\text{I}]/K_I)$

$K'_m = \text{slope} \times V$

$= \text{slope}/(9.3 \text{ min mM}^{-1})$

$[Nbs_2]$ (mM)	Slope (min)	K'_m (mM)
0.25	31.3	3.36
0.5	40.4	4.34
1.0	57.7	6.20

A plot of K'_m vs $[Nbs_2]$ gives

Intercept = $K^R_m = \underline{2.4\ \text{mM}}$

Slope = $K^R_m/K_I = 3.8$

$K_I = (2.4\ \text{mM})/3.8 = \underline{0.63\ \text{mM}}$

$V_R = (9.3 \text{ min mM}^{-1})$

$= 0.1075$ mM min^{-1}

$= \underline{1.79\ \mu\text{M s}^{-1}}$

b) $$K^{eq} = V_F\, K_m^R / V_R K_m$$
$$= (1.17\times10^3)(5\times10^{-7})(2.4\times10^{-3})/[(1.79\times10^{-6})(5.7\times10^{-5})]$$
$$= \underline{1.38\times10^4}$$

CHAPTER 9

1. a) $$\nu_o(K) = \varphi / h = \frac{(2.2eV)(1.6\times10^{-19}\,J/eV)}{6.626\times10^{-34}\,Js}$$

$$\nu_o(K) = \underline{5.31\times10^{14}s^{-1}}$$

$$\lambda_o(K) = c/\upsilon_o = \frac{2.998\times10^{8}\text{m s}^{-1}}{5.31\times10^{14}\text{s}^{-1}} 5.64\times10^{-7}\text{m}$$

$$= \underline{564\text{ nm}}$$

$$\upsilon_o(Ni) = \frac{\phi(Ni)}{\phi(K)}\upsilon_o(K) = \frac{5.0}{2.2}\upsilon_o(K) = \underline{12.09\times10^{14}s^{-1}}$$

$$\lambda_o(Ni) = \frac{\phi(K)}{\phi(Ni)}\lambda_o(K) = \underline{248\text{ nm}}$$

b) A wavelength of 400 nm will eject K electons but not Ni electrons.

c) $$K.E. = h\upsilon - \phi_o(K) = h[\frac{c}{\lambda} - \upsilon_o(K)]$$

$$= 6.626\times10^{-34}(\frac{2.998\times10^{8}}{400\times10^{-9}} - 5.32\times10^{14}) = \underline{1.44\times10^{-19}}J$$

2. a) $$\lambda = \frac{h}{p} = \frac{6.626\times10^{-34}}{(9.109\times10^{-31})(1.5\times10^{8})} = 4.85\times10^{-12}m$$

$$= \underline{4.85\times10^{-3}}\text{nm}$$

b) $$\Delta p = \frac{h}{4\pi}/\Delta x = \frac{6.626\times10^{-34}}{4\pi10^{-10}} = \underline{5.27\times10^{-25}\text{kg m s}^{-1}}$$

3. $$\Delta E = \frac{(n_{x^2} + n_{y^2} + n_{z^2})h^2}{8\text{ ma}^2} = \frac{3h^2}{8\text{ ma}^2}$$

$$\Delta E = \frac{1.64\times10^{-67}}{\text{ma}^2}$$

At 298 K kT = 4.11×10^{-21} J

At 1 K kT = 1.38×10^{-23} J

a) $m = \dfrac{4.003 \text{ g mol}^{-1}}{(1000 \text{ g kg}^{-1})(6.023\times10^{23})} = 6.65\times10^{-27} \text{ kg}$

$a = 10^{-6} \text{ m}$

$\Delta E = 2.47\times10^{-29} \text{ J}$

$\Delta E/kT = 6.0\times10^{-9}$ at 298 K

$\Delta E/kT = 1.8\times10^{-6}$ at 1 K

System is classical.

b) $m = 4.15\times10^{-23}$ kg

$\text{a} = 10^{-8}$ m

$\Delta E = 3.95\times10^{-29}$ J

$\Delta E/kT = 9.6\times10^{-9}$ at 298 K

$\Delta \text{E}/kT = 2.9\times10^{-6}$ at 1 K

System is classical.

c) $\Delta E = 2.47\times10^{-21}$ J

$\Delta E/kT = 0.60$ at 298 K

$\Delta \text{E}/kT = 179$ at 1 K

System is quantum mechanical.

4. a) $E = n^2 h^2/8 \text{ ml}^2; \lambda_{1\to2} = hc/(E_2 - E_1) = 8 \text{ mcl}^2/3 \text{ h}$

$l = 1\times10^{-9} \text{ m}; E_1 = 6.02\times10^{-20} \text{ J}; \lambda_{12} = 11{,}000 \text{ Å}$

b) $l = 2\times10^{-9} \text{ m}; E_1 = \dfrac{6.02\times10^{-20}}{4} = 1.51\times10^{-20} \text{ J}; \lambda_{12} = 44{,}000 \text{Å}$

c) Both electrons are in ground state; energy is twice that in part (a).

$E_1 = 1.204\times10^{-19}$ J. Longest wavelength transition is same as part (a) $\lambda_{12} = \underline{11{,}000}$ Å.

d) $-\dfrac{\hbar^2}{2m_e}\left[\dfrac{d^2\psi_{1n_1}(x)}{dx^2} + \dfrac{d^2\psi_{2n_2}(x)}{dx^2}\right] = E_{1n_1}\psi_{1n_1}(x) + \text{E}_{2n_2}\psi_{2n_2}(x)$

e) Because the electrons repel one another, V is positive and the total energy of each state is *increased* accordingly. The interaction will be greatest when $N_1 = n_2 = 1$, and less when $n_1 = 1$ and $n_2 = 2$. Therefore, E, is increased, but $\Delta E = E_2 - E_1$ is decreased by the interaction.

5. a) $\psi_1^2 = (2/a)\sin^2(\pi x/a)$; $a = 1$ nm

$\text{Probability} = (2/1)\int_{0.49}^{0.51} \sin^2(\pi x/1)\, dx$

Probability = 0.04

A good approximation is $\text{Probability} = \psi_1^2(x = 0.5)\Delta x$

$\text{Probability} \cong 2\sin^2(\pi/2)(0.01) = \underline{0.04}$

b) $\psi_2^2 = (2/a)\sin^2(2\pi x/a)$

Probability $\equiv 2\sin^2(\pi)(0.01) = 0$

c) $\lambda = \dfrac{8mcl^2}{h(2N+1)}$

If the mass doubles, λ doubles: $\lambda = 400$ nm

If the charge doubles, λ is unchanged: $\lambda = 200$ nm

If the length doubles, λ is increased by four: $\lambda = 800$ nm

6. a) It is continuous at the boundaries because $\lim_{x\to 0} x(x-a) = 0$ and $\lim_{x\to a} x(x-a) = 0$

b) $E' = \int_0^a \Psi(x) H\Psi(x)\,dx \Big/ \int_0^a \Psi(x)^2\,dx$

$$\Psi(x)H\Psi(x) = x(x-a)\left[\frac{-h^2 d^2(x^2-ax)}{8\pi^2 m\,dx^2}\right]$$

$$= \frac{-h^2}{4\pi^2 m}(x^2-ax)$$

$$E' = \frac{-h^2}{4\pi^2 m}\int_0^a (x^2-ax)dx \Big/ \int_0^a (x^4-2ax^3+a^2x^2)dx$$

$$= \frac{-h^2}{4\pi^2 m}\left[\frac{a^3}{3}-\frac{a^3}{2}\right] \Big/ \left[\frac{a^5}{5}-\frac{2a^5}{4}+\frac{a^5}{3}\right] = \frac{5h^2}{4\pi^2\,ma^2} = \frac{0.127h^2}{ma^2}$$

c) 1.6% too high.

d) Try $x^2(x-a)^2$, which is both continuous and has a continuous first derivative. Calculate E' and see whether it is lower than that calculated in part (b). If so, it is a better approximation.

7. a) $E_{n_x n_y} = \dfrac{h^2}{8\,ma^2}(n_{x^2} + n_{y^2})$

b) $E_{11} = \dfrac{h^2}{9\,ma^2}(1+1) = 2E_o$

$E_{21} = E_{12} = 5E_o$; $E_{22} = 8E_o$; $E_{13} = E_{31} = 10E_o$;

$E_{23} = E_{32} = 13E_o$; $E_{14} = E_{41} = 17E_o$; $E_{33} = 18E_o$;

$E_{24} = E_{42} = 20E_o$; $E_{34} = E_{43} = 25E_o$.

c) There are two electrons in each of the first $13\,\pi$ orbitals. Therefore, levels through $E_{24} - E_{42}$ are filled.

d) $\lambda_{24\to 34} = \dfrac{8\,mcl^2}{(25-20)h} = 6.60\times 10^{-7}\,m = \underline{660\text{ nm}}$

8. a) $H = -(h^2/8\pi^2 m)\left(\dfrac{d^2}{dx^2}+\dfrac{d^2}{dy^2}+\dfrac{d^2}{dz^2}\right)\psi(x,y,z)$

b) $\psi_{n_1n_2n_3} = (8/a^3)^{1/3}\sin(n_1\pi x/a)\sin(n_2\pi y/a)\sin(n_3\pi z/a)$

$E_{n_1n_2n_3} = h^2(n_1^2+n_2^2+n_3^2)/8\,\mathrm{ma}^2 = (n_1^2+n_2^2+n_3^2)E_o$

c) $E_{111} = 3E_o;\ E_{112} = E_{121} = E_{211} = 6E_o$

$E_{122} = E_{212} = E_{221} = 9E_o, E_{113} = E_{131} = E_{311} = 11E_o.$

$E_{222} = 12E_o$

d) The 20 electrons fill up through $E_{113} = E_{131} = E_{311} = 11E_o$.

e) Transition is from $11E_o$ to $12E_o$.

$E_o = h^2/8\,\mathrm{ma}^2 = 6.681\times10^{-19}$ J

$\Delta E = (12-11)E_o = 6.681\times10^{-19}$ J

$\lambda = hc/\Delta E = 2.97\times10^{-7}$ m

$\lambda = \underline{297\text{ nm}}$

9. $<x> = \int_{-\infty}^{\infty} xe^{-2ax^2}dx \,/ \int_{-\infty}^{\infty} e^{-2ax^2}dx$

As the numerator is an odd function $[f(x) = -f(-x)]$ its integration over even limits is zero.

$<x> = 0$

$$<p> = \frac{h}{2\pi i}\int_{-\infty}^{\infty} e^{-ax^2}\frac{de^{-ax^2}}{dx}dx \Big/ \int_{-\infty}^{\infty} e^{-2ax^2}dx$$

$$= \frac{-2ah}{i}\int_{-\infty}^{\infty} xe^{-2ax^2}dx \Big/ \int_{-\infty}^{\infty} e^{-2ax^2}dx = 0$$

10. a) $\dfrac{1}{\mu} = \dfrac{1}{m_H} + \dfrac{1}{m_F} = \dfrac{1}{1.673\times10^{-27}} + \dfrac{1}{(18.85)(1.673\times10^{-27})}$

$\underline{\mu = 1.589\times10^{-27}\text{ kg}}$

(b) $U = \frac{1}{2}(970\text{ N m}^{-1})(x - 0.92\times10^{-10}\text{ m})^2$

(c) $E_v = h\nu_0(v+\frac{1}{2})$

$$\nu_0 = \frac{1}{2\pi}\left(\frac{k}{\mu}\right)^{1/2} = 1.24\times10^{14}\text{ s}^{-1}$$

$E_o = \frac{1}{2}h\nu_o = 4.12\times10^{-20}$ J

$E_1 = 3E_o;\ E_{2o} = 5E_o;\ E_3 \ldots$

11. a) $$\frac{\nu(C_2H_4)}{\nu(C_2D_4)} = \left[\frac{\mu(C_2D_4)}{\mu(C_2H_4)}\right]^{1/2}$$

11. a) $$\frac{\nu(C_2H_4)}{\nu(C_2D_4)} = \left[\frac{\mu(C_2D_4)}{\mu(C_2H_4)}\right]^{1/2}$$

$$\mu(C_2H_4) = (1/2)(\text{mass of } CH_2)$$

$$\frac{\nu(C_2H_4)}{\nu(C_2D_4)} = \left(\frac{8.0215}{7.0135}\right)^{1/2} = \underline{1.069}$$

b) The vibration frequency will increase for a shorter bond. A short bond means a stronger force constant; ν is proportional to $k^{1/2}$.

c) The first harmonic is $4000\ \text{cm}^{-1}$; $\lambda = (4000)^{-1}$ cm $\lambda = \underline{2500\ \text{nm}}$.

12. a) For $N^{14}O^{16}$

$$m_p = 1.6726\times10^{-27}; m(N^{14}) = 14.00307; m(O^{16}) = 15.99491; m(H) = 1.00797$$

$$\frac{1}{\mu} = \frac{m(H)}{m_p}\left(\frac{1}{m(N^{14})} + \frac{1}{m(O^{16})}\right)$$

$$\mu(N^{14}O^{16}) = 1.239\times10^{-26}\ \text{kg}$$

$$k = 4\pi^2\nu_0^2\mu = 4\pi^2(1904\ \text{cm}^{-1} \bullet 3\times10^{10}\ \text{cm s}^{-1})^2(1.239\times10^{-26}\ \text{kg})$$

$$= \underline{1.596\times10^3\ \text{N m}^{-1}}$$

b) $$m(N^{15}) = 15.00011$$

$$\mu(N^{15}O^{16}) = 1.284\times10^{-26}$$

$$\nu(N^{15}O^{16})/\nu(N^{14}O^{16}) = \sqrt{\mu(N^{14}O^{16})/\mu(N^{14}O^{16})} = 0.9823$$

$$\nu(N^{14}O^{16}) = \underline{1870\ \text{cm}^{-1}}$$

c) $$\mu(N^{14}O^{16}\ \text{bound}) = m_p \cdot m(N^{14})/m(H) = 2.324\times10^{-26}$$

$$\nu(N^{14}O^{16}\ \text{bound}) = \sqrt{(1.239/2.324)}(1904) = \underline{1390\ \text{cm}^{-1}}$$

d) $$k = 4\pi^2\nu_0^2\mu = 4\pi^2(1615.3\times10^{10})^2(1.239\times10^{-26})$$

$$= \underline{1.148\times10^3\ \text{N m}^{-1}}$$

e) Assumption (c):

$$\mu(N^{15}O^{16}\ \text{bound}) = 2.489\times10^{-26}$$

$$\nu(N^{15}O^{16}\ \text{bound}) = \sqrt{(1.284/2.489)}(1870) = \underline{1343\ \text{cm}^{-1}}$$

Assumption (d):

$$\nu = (1/2\pi)\sqrt{k/\mu} = (1/2\pi)\sqrt{1.148\times10^3/1.284\times10^{-26}} = 4.759\times10^{13}\ \text{s}^{-1}$$

$$\nu = 4.759\times10^{13}\ \text{s}^{-1}/3\times10^{10}\ \text{cm s}^{-1} = \underline{1586\ \text{cm}^{-1}}$$

13. $$E_n = -Z^2\,R/n^2$$

$$E_1(He^+) = -4(13.6) = \underline{-54.4\ \text{eV molecule}^{-1}}$$

$$= -4(1312) = \underline{-5248\ \text{kJ mol}^{-1}}$$

Ionization potential = $-E_1$

14. a) $$H = \frac{-h^2}{2m_e}V^2 - \frac{2e^2}{r_1} - \frac{e^2}{r_2} + \frac{2e^2}{d}$$

We can use m_e instead of μ, because we have assumed the nuclei to be stationary.

b) $$\frac{-\hbar^2}{2m_e}V^2\psi_n + \left(\frac{2e^2}{d} - \frac{2e^2}{r_1} - \frac{e^2}{r_2}\right)\psi_n = E_n\psi_n$$

The subscript *n* represents 3 quantum numbers. We choose a coordinate system which is convenient for this problem (probably elliptic coordinates) and write V, ψ_n, r_1 and r_2 in terms of these coordinates.

c) Calculate the energy as a function of d and see if any of the states show a minimum.

15. a)

	NO		NO⁻		NO⁺	
$\sigma^*(2p)$	___		___		___	
$\pi^*(2p)$	X	___	X	X	___	___
$\pi(2p)$	XX	XX	XX	XX	XX	XX
$\sigma(2p)$	XX		XX		XX	
$\sigma^*(2s)$	XX		XX		XX	
$\sigma(2s)$	XX		XX		XX	
$\sigma^*(1s)$	XX		XX		XX	
$\sigma(1s)$	XX		XX		XX	

NO NO$^-$ NO$^+$

b) Bond energies $NO^- < NO < NO^+$
Bond lengths $NO^+ < NO < NO^-$

16. a) $\int_0^R [\psi(1\,s)]^2 dv = 0.90$; solve for R

From Example 9.5 integral is equal to:

$$\frac{-e^{-2\rho}}{2}(4\rho^2 + 4\rho + 2)\Big|_0^{R/a_0} \quad a_0 = \text{Bohr radius}$$

$$\frac{-e^{-2R/a_0}}{2}[4(R/a_0)^2 + 4(R/a_0) + 2] = 0.90$$

$R \to \infty$ for 100% probability

b) From Eq. 9.32 $\Delta E = (\frac{3}{4})2.179\times10^{-18}$ J molecule^{-1}

$\Delta E = 1.634\times10^{-18}$ J molecule^{-1}

$$\lambda = \frac{hc}{\Delta E} = \frac{(6.626\times10^{-34}\text{ J s})(2.998\times10^{8}\text{ m s}^{-1})}{(1.634\times10^{-18}\text{ J})}$$

$\lambda = 1.216\times10^{-7} m = 121.6\text{ nm} = 1216\text{Å}$

Ultraviolet wavelengths

c) $$\frac{-\hbar^2}{2m}\left(\frac{\partial^2\psi}{\partial x^2}+\frac{\partial^2\psi}{\partial y^2}+\frac{\partial^2\psi}{\partial t^2}\right)+U(x,y,z)\psi = E\psi$$

$U(x,y,z)=0$ when $0<x<a,\ 0<y<a,\ 0<z<a$

$U(x,y,z)=\infty$ when $x\le 0,\ x\ge a, y\le 0, y\ge a, z\le 0, z\ge a$

d) $$\Delta E = \frac{3h^2}{8\text{ ma}^2} = \frac{3(6.626\times10^{-34})^2}{(8)(1.67\times10^{-27})(10^{-2})^2}$$

$\Delta E = 9.86\times10^{-37}$ J

$$\lambda = \frac{hc}{\Delta E} = 2\times10^{11}\text{ m}$$

Radiowaves can be km in length, so these wavelengths are ultralong radiowaves.

17. a) Choose the $C-C$ bond of acetylene along z. Each carbon atom has an sp hybrid along z, and two p orbitals (one along x and one along y).

b) The $x-y$ plane is the molecular plane for ethylene. Each carbon has an sp^2 hybrid in $x-y$ plane, and a P_z orbital.

c) Each carbon has an sp^3 orbital.

18. π-type orbitals constructed from $O(2p)+C(2p_z)+N(2p_z)$ contain electrons involved in bonding. Because these are perpendicular to the OCN plane, the structure shown becomes planar. sp^2 hybrid orbitals involved in σ bonds predicts 120° bond angles. Steric repulsions and electrostatic interactions will modify these by a few degrees.

19. (a)

```
                              H
                              |
C(1)H3 - C(2)H = C(3)H - C(4) = O
```

C(1) sp^3

C(2) sp^2 (x, y plane)

C(3) sp^2 (x, y plane)

C(4) sp^2 (x, y plane)

O sp^2 (x, y plane)

b) The four π MO's are linear combinations of four p_z atomic orbitals on C(2), C(3), C(4) and O. The phases and ordering are the same as Fig. 9.24.

c) All the atoms in crotonaldehyde are in a plane, except for the three protons of C(1) H_3. There is free rotation around the C(1) - C(2) bond.

d) The N and the R' atom bonded to the N will be in the plane formed by RCH. The $RC=N-CR'$ bond should be 120°.

e) The lone pair on the N is protonated to give a σ bond which is part of an sp^2 hybrid.

20. a) Each carbon atom contributes 1 π electron. N(1), N(3) and N(7) only have 2σ bonds; they thus contribute 1 π electron (the other two are in a nonbonding orbital). N(9) and N(10) have $3\ \sigma$; they each contribute 2π electrons. The total number of π electrons is 12/32. As a check make sure that all electrons are counted: 20 core, σ (16 σ bonds), 12π and 6 nonbonding.

b) Each carbon and nitrogen have a p_z orbital in the π system. There are 10 p_z atomic orbitals; and thus $\underline{10\pi \text{ and } \pi^* \text{ MOs}}$.

c) Each carbon and nitrogen have one 2*s* and two 2*p* atomic orbitals to contribute. Therefore there are 30 σ, σ^* and *n* orbitals which can be formed from 2*s* and 2*p* atomic orbitals.

d) N(1), N(3), N(7).

21. One can plot the data and use trigonometry to obtain

a) Ground state $\begin{cases} \mu = 1.95\,\text{D} \\ \mu_\perp = 1.70\,\text{D} \end{cases}$

Excited state $\begin{cases} \mu = 0.28\,\text{D} \\ \mu_\perp = 0.41\,\text{D} \end{cases}$

b) $|\mu|_G = 2.95\,\text{D}; |\mu|_E = 0.50\,\text{D}; \Delta|\mu| = -2.09\,\text{D}$

c) $\theta = -41.2°$ from CN direction (roughly opposite to $C=O$ bond direction)
$\theta_E = 10.9°$ (nearly parallel to CN bond, but displaced toward $C=O$)

22. a)

$$\Psi_1 = C_{11}\Phi_O + C_{12}\Phi_C + C_{13}\Phi_N$$
$$\Psi_2 = C_{21}\Phi_O + C_{22}\Phi_C + C_{23}\Phi_N$$
$$\Psi_3 = C_{31}\Phi_O + C_{32}\Phi_C + C_{33}\Phi_N$$

In analogy with the particle-in-a-box, C_{11}, C_{12}, C_{13} are all positive; $C_{22} = 0$, C_{21} is positive, and C_{23} is negative; C_{31} and C_{33} are positive and C_{32} is negative.

b)

$$\begin{matrix} C_{11} & C_{12} & C_{13} \\ C_{21} & C_{22} & C_{23} \\ C_{31} & C_{32} & C_{33} \end{matrix} = \begin{matrix} 1/2 & 1/\sqrt{2} & 1/2 \\ 1/\sqrt{2} & 0 & -1/\sqrt{2} \\ 1/2 & -1/\sqrt{2} & 1/2 \end{matrix}$$

The above coefficients fit all the requirements for normalization and orthogonality.

c) The order is ψ_1 with the lowest energy and ψ_3 the highest. O and C contribute one π electron each; N contributes two π electrons. The first two π MOs are filled with the four electrons.

d) The electron density is proportional to ψ^2.

$$\Psi_1^2 = \tfrac{1}{4}\phi_{\mathrm{O}}^2 + \tfrac{1}{2}\phi_{\mathrm{C}}^2 + \tfrac{1}{4}\phi_{\mathrm{N}}^2$$

$$\Psi_2^2 = \tfrac{1}{2}\phi_{\mathrm{O}}^2 + \tfrac{1}{2}\phi_{\mathrm{N}}^2$$

We assume no overlap between P_z orbitals on different atoms.

e) The integral of ϕ_{O}^2 or ϕ_{C}^2 or ϕ_{N}^2 overall space $=1$, therefore sum of squares of coefficients $= 1 =$ integral of ψ's overall space.

f) π charge on O $= -2(1/4 + 1/2) = 3/2$

π charge on C $= -2(1/2) = -1$

π charge on N $= -3/2$

g) Net charge on O $= -3/2 + 1 = -1/2$

Net charge on C $= -1 + 1 = 0$

Net charge on N $= -3/2 + 2 = 1/2$

$$\mu_x \Sigma x_i q_i$$

$$\mu_x = (0)(1) + (1.34)(1/2) + (-0.62)(-1/2)$$

$$\mu_x = 0.98 \text{ electron Å}$$

$$\mu_y = (0)(1) + (0)(1/2) + (-1.07)(-1/2)$$

$$\mu_y = 0.335 \text{ electron Å}$$

$$1 \text{ electron Å} = 4.8 \times 10^{-18} \text{ esu cm} = 4.8 \text{ D}$$

$$\mu_x = 4.70 \text{ D}$$

$$\mu_y = 2.57 \text{ D}$$

23. a) $$U = \frac{1389(+1)(-1)}{0.529} = \underline{-2626 \text{ kJ mol}^{-1}}$$

$$U = \frac{1389(+0.417)(-0.834)}{1.5} = \underline{-322 \text{ kJ mol}^{-1}}$$

b) Torsion angle rotation changes most easily and bond length changes the least. The potential energy functions given in Eqs. 9.5, 9.6, 9.7 characterize how the energy changes with variations in bond lengths, bond angles and torsion angles.

24. a) $U = k_r(r - r_{eq})^2$

$= (1200 \text{ kJ mol}^{-1}\text{Å}^{-2})(1.750 \text{ Å} - 1.540 \text{ Å})^2$

$U = \underline{52.9 \text{ kJ mol}^{-1}}$

b) $U = k_b(\theta - \theta_{eq})^2$

$= (0.0612 \text{ kJ mol}^{-1} \text{ deg}^{-2})(109.5 \text{ deg} - 90.0 \text{ deg})^2$

$U = \underline{23.3 \text{ kJ mol}^{-1}}$

c) $U = k[1 + \cos(3\phi)]$

$\frac{dU}{d\phi} = 0$ at maxima and minima

$\sin 3\phi = 0$ and therefore $3\phi = n\pi (n = \text{integer})$ at maxima and minima

$\phi = 0, \frac{2\pi}{3}, \frac{4\pi}{3}, \ldots = 0°, 120°, 240°, 360° = \text{maxima}$

$\phi = \frac{\pi}{3}, \frac{3\pi}{3}, \frac{5\pi}{3} \ldots = 60°, 180° \; 300° = \text{minima}$

CHAPTER 10

1. a) $E = \frac{hc}{\lambda} = \frac{(6.626\times10^{-34}\text{ J s})(3\times10^{8}\text{ m s}^{-1})}{6.8\times10^{-7}\text{ m}} = \underline{2.92\times10^{-19}\text{ J}}$

   $= 2.92\times10^{-19}/1.602\times10^{-19} = \underline{1.82\text{ eV}}$

   $= (2.92\times10^{-19})(6.024\times10^{20}) = \underline{175.9\text{ kJ mol}^{-1}}$

   b) Einsteins $= 485/175.9 = \underline{\text{kJ mol}^{-1}}$

   c) $\phi = 2.76/8$ to $2.76/9 = \underline{0.345\text{ to }0.307}$

2. Ruby laser produces $1\times10^{12}\ \mu\text{W cm}^{-2}$ for 1 ms. Assume all the energy is absorbed by the eye.

   Energy $= (10^{6}\text{ W})(10^{-3}\text{ s}) = 10^{3}$ joule $= 239$ cal

   $$\text{Temperature rise} = \Delta T = \frac{\text{heat}}{(C_p)(\text{weight})} = \frac{239\text{ cal}}{(1\text{ cal g}^{-1}\text{ deg}^{-1})(10^{-2}\text{ g})}$$

   $$\Delta T = 2.4\times10^{4}\text{ deg.}$$

   To make $\Delta T < 2.4$ deg, use absorbance $= 4$, because $I = I_o 10^{-A}$

3. a) Sunburning radiation $= (0.002)\left(\frac{2100\text{ J}}{\text{cm}^2 d}\right)\left(\frac{d}{12\text{ h}}\right)\left(\frac{h}{3600\text{ s}}\right)$

   $= \underline{9.72\times10^{-5}\text{ J cm}^{-2}\text{ s}^{-1}}$

   b) The benzoates absorb the incident light; the TiO_2 reflects the incident light.

4. a) $c = A/\varepsilon l = 1/(100)(1) = \underline{10^{-2}\text{ M}}$

   b) $A = \log(I_o/I) = -\log 0.01 = 2$

   $c = \underline{2\times10^{-2}\text{ M}}$

   c) Concentration of pure benzene $= \left(\frac{0.8\text{ g}}{\text{cm}^2}\right)\left(\frac{1000\text{ cm}^3}{l}\right)\left(\frac{\text{mole}}{7.8\text{ g}}\right)$

   $c = 10.3$ M

   $l = A/\varepsilon c = 1/(100)(10.3) = \underline{9.8\times10^{-4}\text{ cm}}$

5. Use Eq. (10.12) to obtain

   $(\text{In}^-) = 1.73\times10^{-3}$; $(\text{HIn}) = 3.33\times10^{-5}$

   $(\text{In}^-/\text{HIn}) = 52$

   $pH = pK + \log(\text{In}^-/\text{HIn}) = 5 + 1.7 = \underline{6.7}$

6. Use Eq. (10.12) to obtain

$$c_a(\mathrm{g\,l^{-1}}) = \frac{9.27\,A_{645} - 45.60\,A_{663}}{(16.75)(9.27) - (82.04)(45.60)}$$

$$= \frac{9.27A_{645} - 45.60\,A_{663}}{-3586}$$

$$c_a(\mu\mathrm{g\ ml^{-1}}) = 12.80A_{663} - 2.58A_{645}$$

$$c_b(\mu\mathrm{g\ ml^{-1}}) = 22.88\,A_{645} - 4.67\,A_{663}$$

$$c_{a+b}(\mu\mathrm{g\ ml^{-1}}) = 8.13A_{663} + 20.30\,A_{645}$$

7. a) $A = (\varepsilon_p[\mathrm{P}] + \varepsilon_c[\mathrm{C}])1$

$[\mathrm{P}] = P_o - [\mathrm{C}]1 = 1$

$A = \varepsilon_p P_o + (\varepsilon_C - \varepsilon_P)[\mathrm{C}]$

$$[\mathrm{C}] = \frac{A - \varepsilon_p P_o}{\varepsilon_c - \varepsilon_p} = \frac{1.4 - (12{,}000)(1\times10^{-4})}{3000}$$

$[\mathrm{C}] = \underline{6.67\times10^{-5}\ \mathrm{M}}$

b) $$K = \frac{[\mathrm{C}]}{[\mathrm{P}][\mathrm{D}]} = \frac{6.67\times10^{-5}}{(3.33\times10^{-5})(193.3\times10^{-5})}$$

$K = \underline{1.04\times10^{3}\ \mathrm{M^{-1}}}$

c) The CD at 208 nm is a measure of the α-helix content of a protein as shown in Fig. 10.24.

8. a) $$\mathrm{InH^+} = \frac{A_{400}}{\varepsilon_{400}} = \frac{0.25}{10^4} = 2.5\times10^{-5}\ \mathrm{M}$$

$$\mathrm{In} = \frac{A_{460}}{\varepsilon_{460}} = \frac{0.300}{1.2\times10^4} = 2.5\times10^{-5}\ \mathrm{M}$$

$pH = pK = \underline{4}$

b) $\lambda = 440$ is an isosbestic, therefore $A = 0.400$ is independent of pH. ε_c

c) Excite each species at a wavelength where only it absorbs. Excite $\mathrm{InH^+}$ at $\underline{\lambda = 400\ \mathrm{nm}}$; In at $\underline{\lambda = 460\ \mathrm{nm}}$.

9. a) $$\frac{-dN}{dt} = k\,N_0 = \left(\frac{\ln 2}{t_{1/2}}\right)N_0 = \frac{(\ln 2)(1000)}{(14.3\ \mathrm{d})(86{,}400\ \mathrm{s\ d^{-1}})}$$

$$\frac{-dN}{dt} = 5.61\times10^{-4}\,s^{-1}$$

Maximum counts = $\underline{100}$

b) $$\varepsilon = \frac{(70{,}000\ \mathrm{mol^{-1}\,L\,cm^{-1}})(1000\ \mathrm{cm^3\,L^{-1}})}{(6.023\times10^{23}\ \mathrm{mol^{-1}})}$$

$\varepsilon = 1.16\times10^{-16}\ \mathrm{cm}^2$ = photon capture cross-section

c) $$I_0 = \frac{2\times10^{-3}\ \text{J s}^{-1}\ \text{cm}^{-2}}{(\text{hc}/\lambda)} = \frac{(2\times10^{-3}\ \text{J s}^{-3}\ \text{cm}^{-2})(485\ \times10^{-9}\ \text{m})}{(6.626\times10^{-34}\ \text{J s})(3\times10^{8}\ \text{m s}^{-1})}$$

$$I_0 = 4.9\times10^{15}\text{s}^{-1}\ \text{cm}^{-2}$$

$$I_F = f_F I_{\text{absorbed}} = f_F \varepsilon I_0 N_0$$

$$= (0.93)(1.1610^{-6}\ \text{cm}^2)(4.910^{15}\ \text{s}^{-1}\ \text{cm}^{-2})(1000)$$

$$I_F = \underline{530\ \text{photons s}^{-1}}$$

d) $$\text{Maximum photons} = \left(\frac{f_F}{f_{\text{degredation}}}\right)N_0 = \frac{(0.93)(1000)}{10^{-6}}$$

$$= \underline{9.3\times10^{8}}$$

e) Based on counting rates fluorescence is more sensitive than radioactivity, because each molecule can fluoresce many times but radioactively decay only once. However background signals are very important. Everything fluoresces, so it is difficult to separate the analyte photons from solvent, impurities, etc.

10. a) Plot ln (intensity) *vs*. time; slope is reciprocal of lifetime. Alternatively, use linear regression function on many pocket calculators to obtain slope. $\underline{\tau = 6.9\ \text{ns}}$

b) $k_o = (0.7)/(6.9) = \underline{1.0\times10^{8}\text{s}}$

c) $\text{Eff} = 1 - (\tau_{D+A})/(\tau_D) = 1 - (6/6.9)$
$\text{Eff} = 0.565$

$$r = r_o(1/\text{Eff} - 1)^{1/6}$$

$$r = 20(0.957)$$

$$\underline{r = 19\,\text{Å}}$$

11. a) Choose an isosbestic point $(\lambda = 273\ \text{nm},\ 284\ \text{nm})$. At both these wavelengths the extinction coefficient is $1000\ \text{cm}^{-1}\text{M}^{-1}$ (per mole of tyrosine). At these wavelengths $A = 0.6$.
$c = A/\varepsilon = 6.0\times10^{-4}$ M tyrosine

$c = \underline{2\times10^{-4}\ \text{M peptide}}$

b) At 300 nm only tyrosinate ion absorbs ($\varepsilon = 1500$).

$$c = A/\varepsilon = 0.2/1500 = 1.33 \times 10^{-4}\,\mathrm{M}$$

$$= \mathrm{TYR^{-}TYR} + \mathrm{TYR^{-}} = 6 \times 10^{-4}$$

$$\mathrm{TYR} = 4.67 \times 10^{-4}$$

$$K = \frac{[\mathrm{TYR^{-}}][\mathrm{H^{-}}]}{[\mathrm{TYR}]}$$

$$[\mathrm{H^{+}}] = \frac{(4.67 \times 10^{-4})(10^{-10})}{(1.33 \times 10^{-4})}$$

$$[\mathrm{H^{+}}] = 3.51 \times 10^{-10}$$

$$\underline{\mathrm{pH} = 9.45}$$

12. a) $A = \varepsilon cl = (10^4)(10^{-2})(1) = 10^2$. All the incident photons are absorbed. Quantum yield is 1, so 10^{15} photons s^{-1} are fluoresced.

b) Fluorescence occurs in all directions; the detector only senses the fraction that strike it.

c) Phosphorescence occurs at longer wavelengths. Triplet excited states have less electron-electron repulsion than the corresponding singlet excited state.

d) When light is absorbed, energy is of course conserved. When one of the paths for energy utilization is blocked, there is more energy available for the other paths such as fluorescence.

13.
$$K = \frac{\alpha_2}{[\alpha]^2}$$

$$A = \varepsilon_{\alpha_2}[\alpha_2] + \varepsilon_{\alpha}[\alpha];\ l = 1$$

$$[\alpha] + 2[\alpha_2] = \alpha_0 = \text{total concentration of protein}$$

$$A = \varepsilon_{\alpha_2}[\alpha_2] + \varepsilon_{\alpha}(\alpha_0 - 2[\alpha_2])$$

$$[\alpha_2] = \frac{A - \varepsilon_{\alpha}\alpha_0}{\varepsilon_{\alpha_2} - 2\varepsilon_{\alpha}}$$

If the molar extinction coefficient of the monomer and dimer are known, the concentration of dimer is easily obtained from the absorbance and the total concentration of protein. If they are not known extrapolation of absorbance to very low or very high concentrations can lead in principle to these values.

14. By measuring the absorbance of mixtures of A and B for different initial concentrations the two unknowns K and ε_c an be determined. The equations we use are:

$$K = \frac{[\mathrm{C}]}{[\mathrm{A}][\mathrm{B}]}$$

$$A = (\varepsilon_A[\mathrm{A}] + \varepsilon_B[\mathrm{B}] + \varepsilon_C[\mathrm{C}])l$$

We simplify the notation by choosing equal initial amounts of A and B.
$[\mathrm{A_o}] = [\mathrm{B_o}] = a_o$

Let $X = [\mathrm{C}]$ = concentration of C at equilibrium. Then,

$$K = \frac{X}{(a_o - X)^2} \tag{1}$$

$$A - (\varepsilon_A + \varepsilon_B)a_o l = (\varepsilon_C - a_A - \varepsilon_B)lX \tag{2}$$

We can solve for X in one equation and substitute into the other to obtain a complicated equation. A non-linear least squares fit to A *vs.* a_o would give K and ε_c. The usual method uses the approximation that $a_o^2 >> X^2$:

$$K = \frac{X}{(a_o^2 - 2a_oX + X^2)} \cong \frac{X}{a_o^2 - 2a_oX}$$

$$X = \frac{Ka_o 2}{1 + 2Ka_o}$$

Now X is substituted into the Eq. (2), and the reciprocal of both sides of the equation is taken. Rearranging gives:

$$\frac{a_o l}{A - (\varepsilon_A + \varepsilon_B)a_o l} = \frac{1}{(\varepsilon_C - \varepsilon_A - \varepsilon_B)K}\left(\frac{1}{a_o}\right) + \frac{1}{(\varepsilon_C - \varepsilon_A - \varepsilon_B)}$$

The right hand side is plotted *vs.* $(1/a_o)$ to obtain
slope $= [K(\varepsilon_C - \varepsilon_A - \varepsilon_B)]^{-1}$; intercept $= (\varepsilon_C - \varepsilon_A - \varepsilon_B)^{-1}$. This is the Benesi-Hildebrand equation. If a_o^2 is not $>> X^2$, successive approximations can be used. A value of ε_C is assumed and Eq. (2) is used to calculate X for different values of a_o. Equation (1) is used to calculate K for each a_o and X. Vary ε_C until the most constant value of K is obtained.

15. a) $\varepsilon_L - \varepsilon_R = (A_L - A_R)/lc = (-0.50\times10^{-4})/(1.00\times10^{-4}) = \underline{-0.50\ \text{M}^{-1}\,\text{cm}^{-1}}$

b) RNA: $\varepsilon_L - \varepsilon_R(260) = (6.00\times10^{-4})(1.00\times10^{-4}) = 6.00\ \text{M}^{-1}\,\text{cm}^{-1}$

$c = (A_L - A_R)/l(\varepsilon_L - \varepsilon_R) = (3.12\times10^{-4})/6.00$

$\underline{c = 0.52\times10^{-4}\ \text{M}}$

DNA: $\varepsilon_L - \varepsilon_R(240) = (-2.20\times10^{-4})/(1.00\times10^{-4}) = -2.20\ \text{M}^{-1}\,\text{cm}^{-1}$

$$c = \frac{-0.24\times10^{-4}}{-2.20} = \underline{0.11\times10^{-4}\ \text{M}}$$

c) The circular dichroism (CD) at 260 nm will decrease according to Fig. 10.22.

d) The absorbance at 260 nm will increase according to Fig. 10.9.

e) The ORD will decrease proportional to the decrease in CD.

16. a) Phenylalanine has an asymmetric carbon.

b) Phenylalanine for the same reason.

c) Phenylalanine and adenine absorb at wavelengths above 250 nm.

d) Measurement of the lifetime would give a good indication, however an absolute determination requires measurement of the spin of the excited state.

17. $\text{Eff} = (\tau_D - \tau_{D+A})/\tau_D = (21-15)/21 = 0.286$

$\text{Eff} = r_o^6/(r_o^6 + r^6); \quad r_o = 3.46\ \text{nm}$

$r^6 = r_o^6(1/\text{E} - 1)$

$r = r_o(2.5)^{1/6} = (3.45)(1.165) = \underline{4\ \text{nm}}$

18. a) $$[\alpha] = \frac{100[\phi]}{M} = \frac{(100)(7500)}{323.2} = \underline{2320\ \text{deg cm}^3\text{dm}^{-1}\,\text{g}^{-1}}$$

$[\theta] = 3298(\varepsilon_L - \varepsilon_R) = \underline{9894\ \text{deg M}^{-1}\,\text{cm}^{-1}}$

b) $A = \varepsilon cl = (8000)(10^{-4}) = \underline{0.8}$

$A_L - A_R = (3)(10^{-4}) = 3\times10^{-4}$

$\alpha = [\alpha]dc$ with c in g cm^{-3}

$$c \text{ in g cm}^{-3} = \left(10^{-4}\frac{\text{mole}}{l}\right)\left(\frac{l}{10^3\text{ cm}}\right)\left(\frac{323.2\text{ g}}{\text{mole}}\right)$$

$$= 3.232\times10^{-5}$$

$$\alpha = (2320)(0.1)(3.232\times10^{-5}) = \underline{7.5\times10^{-3}\text{ deg}}$$

$$\text{or } \alpha = [\phi]lm/100 = (7500)(1)(10^{-4})/100 = \underline{7.5\times10^{-3}\text{ deg}}$$

c) $$\alpha = \frac{(7.5\times10^{-3})(2\pi)}{360} = \underline{1.31\times10^{-4}\text{ rad cm}^{-1}}$$

$$n_L - n_R = \frac{\lambda\alpha}{\pi} = \frac{(2.8\times10^{-5})(1.31\times10^{-4})}{\pi} = 1.16\times10^{-9}$$

$$\phi = \frac{[\theta]\ln(2\pi)}{(360)(100)} = \frac{(9894)(10^{-4})(\pi)}{1800} = \underline{1.73\times10^{-4}\text{ rad cm}^{-1}}$$

19. $k_D = 1/\tau = 1/29 = 3.45\times10^7\text{ s}^{-1}$

$k_D + k_Q(Q) = 1/5.7 = 1.75\times10^8\text{ s}^{-1}$

$$k_Q = \frac{1.75\times10^8 - 3.45\times10^7}{7.2\times10^{-3}} = \underline{1.96\times10^{10}\text{ M}^{-1}\text{ s}^{-1}}$$

$\phi = \tau_Q/\tau_D = 5.7/29 = 0.2$

Relative fluorescence intensity = $\underline{20}$

20. a) $$K = \frac{[S][R]}{[D]}$$

$$C_0 = \text{total concentration of strands} = 200\ \mu\text{m}$$

$$C_0 = 2[\text{D}] + [\text{S}] + [\text{R}]$$

but $[\text{S}] = [\text{R}]$

$$C_0 = 2[\text{D}] + 2[\text{S}]$$

$$\text{f} = \text{fraction in single strands} = \frac{2[\text{S}]}{C_0}$$

$$1 - \text{f} = \text{fraction in double strands} = \frac{2[\text{D}]}{C_0}$$

$$K = \frac{f^2 C_0^{\ 2}}{4(C_0/2)(1-f)} = \frac{f^2 C_0}{2(1-f)}$$

$$A = \varepsilon_D[\text{D}] + S[\text{S}] + R[\text{R}] = \varepsilon_D[\text{D}] + (\varepsilon_S + \varepsilon_R)[\text{S}]$$

$$A_{60} = A(60°\text{C}) = (\varepsilon_S + \varepsilon_R)(C_0/2)$$

$$A_{10} = A(10°\text{C}) = \varepsilon_D(C_0/2)$$

$$A = A_{10}(1-f) + A_{60}f$$

$$f = \frac{A - A_{10}}{A_{60} - A_{10}}$$

Temperature (°C)	f	K(μM)
10		0.0000
15	0.0309	0.098
20	0.0710	0.542
25	0.0988	1.082
30	0.1914	4.528
35	0.3086	13.78
40	0.4630	39.91
45	0.5586	70.71
50	0.6173	99.56
55	0.6574	126.15
60	1.0000	

b) Tm is temperature for $f = 0.5$

$Tm = 41.9°\text{C}$

c) Least squares fit to $\ln K$ vs. $(1/T)$ gives $\ln K = 45.06 - 1.747\times10^4(1/T)$ K in M

$\Delta H^\circ = \underline{145.2\ \text{kJ mol}^{-1}}$

$\Delta S^\circ = \underline{374.6\ \text{J K}^{-1}\ \text{mol}^{-1}}$

$\Delta G^\circ = \Delta H^\circ - T\Delta S^\circ = 33.6\ \text{kJ mol}^{-1}$

21. a) $A = \varepsilon cl = (3\times10^4\ \text{M}^{-1}\text{cm}^{-1})\ (0.10\ \text{M})\ (1\ \text{cm}) = 3000$
An absorbance of 3000 is about three orders of magnitude too large to measure.

b) $\Delta A = \Delta\varepsilon cl = (2.5\ \text{M}^{-1}\text{cm}^{-1})(0.10\ \text{M})\ (1\ \text{cm}) = 0.25$
An absorbance, or difference in absorbance, of 0.25 is easily measured.

c) The fluorescence could be measured.

d) The circular dichroism has the largest % change with temperature, so it is the most sensitive method to use to obtain K.

e) $K = \dfrac{[\text{U}]}{[\text{S}]} = \dfrac{f}{1-f}$

f) $f = \dfrac{P(T) - P_s}{P_U - P_s}$ $\qquad 1 - f = \dfrac{P_U - P(T)}{P_U - P_s}$

$K = \dfrac{P(T) - P_s}{P_U - P(T)}$

22. a) l_1 (c); l_2, l_3 (b); l_4, l_5, l_6 (a)

b) Chemical shift $= \dfrac{220}{100}(200) = 440\ \text{Hz}$

c) Spin-spin splitting of proton (b) = that of proton (a) independent of field.

d) One line halfway between l_2 and l_3.

23. a)

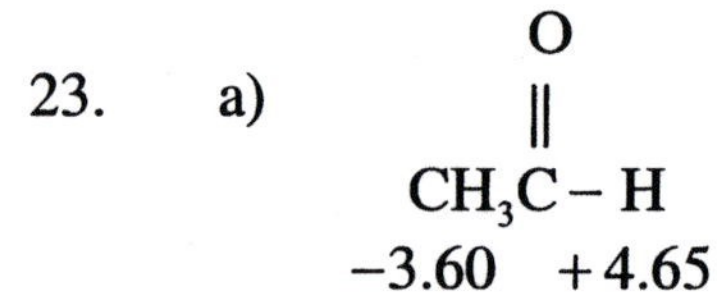

b)

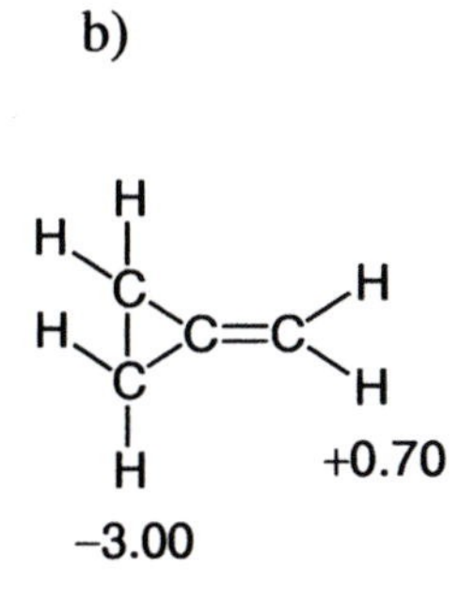

24. a)

$CH_3CH_2CH_2CH_2Cl$ $CH_3CH_2CH(Cl)CH_3$

(a) (b) (c) (d) (a) (b) (c) (d)

As the top spectrum has relative intensities 1:2:3:3, it is 2-chlorobutane. The relative intensities also indicate that (c) resonates at 4 ppm and (b) is near 2 ppm. The spin-spin splittings are consistent with this assignment. They also show that the triplet at high field is (a) and the doublet is (d).

Relative intensities are sufficient to assign the lower spectrum as (d) at 3.5 ppm; (a) at 1 ppm and (b) and (c) in between.

25. a, b) The centers of the multiplets in Hz depend on the frequency of the spectrometer.

Proton.	δ (ppm)	Hz at 500 MHz	Hz at 600 MHz
H1′	6.0	3000	3600
H2′	2.0	1000	1200
H2″	2.5	1250	1500
H3′	4.5	2250	2700
H4′	4.0	2000	2400
H5′	3.8	1900	2280
H5″	3.8	1900	2280

The multiplet patterns are independent of the spectrometer.

H1′: A triplet of intensity 1:2:1 with splitting = 4 Hz.

H2′ : Two triplets each of intensity 1:2:1 with splitting = 4 Hz. The two triplets are split by 10 Hz.

H2″ : The same pattern as for H2′.

H3′ : A quartet of intensity 1:3:3:1 with splitting = 4 Hz.

H4′ : The same pattern as for H3′.

H5′ : A doublet of intensity 1:1 with splitting = 4 Hz.

H5″ : The same pattern as for H5′.

As H5′ and H5″ have identical chemical shifts the intensity of the multiplet at the H5′,H5″ position integrates to two protons.

26. a) A-form RNA

H1′ : A singlet.

H2′ : A doublet of intensity 1:1 with splitting = 5 Hz.

H3′ : A doublet of doublets of intensity 1:1:1:1 with splittings = 5Hz and 9 Hz.

H4′ : A doublet of intensity 1:1 with splitting = 9 Hz. b) B-form RNA

H1′ : A doublet of intensity 1:1 with splitting = 8 Hz.

H2′ : A doublet of doublets of intensity 1:1:1:1 with splittings = 5 Hz and 8 Hz.

H3′ : A doublet of intensity 1:1 with splitting = 5 Hz.

H4′ : A singlet.

c) $\cos^2\phi = (J_{1'2'} - 0.9)/7.7 = 0.9221$; $\cos\phi = \pm 0.9603$ $\phi = \pm 16.2°$, $(180° \pm 16.2°)$

27. a) From Fig. 10.27 we see that the methyl protons, *c*, will resonate farthest upfield. The methylene protons, *a*, will be farthest downfield.

b) The ratio of peak intensities will be $a{:}b{:}c = 2{:}2{:}3$.

c) H_a: A triplet of quadruplets. The splittings for the triplet are 10 Hz; for the quadruplets they are 1 Hz.

H_b: A triplet of quadruplets. The splittings for the triplet are 10 Hz; for the quadruplets they are 3 Hz.

H_c: A triplet of triplets. One splitting is 3 Hz; the other is 1 Hz.

The multiplet intensities are 1:2:1 for the triplets and 1:3:3:1 for the quartet.

28. Glycine: One peak with intensity equal to two protons at $2-3$ ppm. (Fig. 10.27)

Phenylalanine:

$H_a = H_e$ at $7-8$ ppm. A doublet with splitting $J_{ab} = J_{de}$.

$H_b = H_d$ at $7-8$ ppm. A doublet of doublets with splittings of $J_{ab} = J_{de}$ and $J_{bc} = J_{dc}$.

H_c at $7-8$ ppm. A triplet with splitting $J_{bc} = J_{dc}$.

$H_f = H_j$ at $2-3$ ppm. A doublet with splitting of $J_{fg} = J_{jg}$. H_g at $2-3$ ppm. A triplet with splitting $J_{fg} = J_{jg}$.

Intensities are two protons for $H_a = H_e, H_b = H_d, H_f = H_j$.

Intensities are one proton for H_c, H_g.

29. a) For MRI a magnetic field gradient is used. This means that the resonance frequency of water, for example, depends on its position. Thus the resonance frequency of the water reveals it location. A cross section of the distribution of water along the direction of the field gradient is obtained.

b)

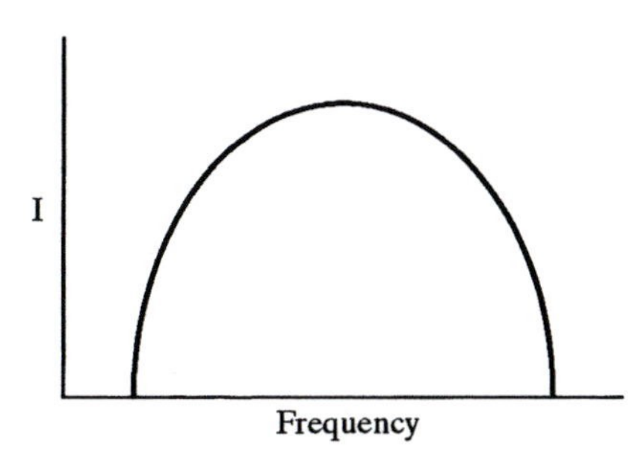

30. a) See section on nuclear Overhauser effect and Example 10.3.

b) The NOE depends on the inverse sixth power of the distance between two protons. Two protons separated by more than 5 Å do not have measurable NOEs by standard methods.

c)

Proton	NOE	α- Helix (Å)
NH - NH	strong	2.8
αH - NH	moderate	3.5
αH - NH_{i+3}	moderate	3.4

Polypeptide is α- helix.

31. a) The cross peaks are symmetrical relative to the diagonal; they have characteristic multiplet patterns. Each cross peak corresponds to the active coupling of two protons; the peaks are further split by the passive coupling of other protons. We use first-order theory and assume all subpeaks are resolved; actual peak widths means that most of the subpeaks are not resolved.

Leucine

$$\begin{array}{ccccccc} & \delta CH_3 & & H\beta & & H\alpha & \\ & | & & | & & | & \\ H\gamma - & C & - & C & - & C & - COO^- \\ & | & & | & & | & \\ & \delta' CH_3 & & H\beta' & & ND_3 + & \end{array}$$

H_α - H_β crosspeak (32 subpeaks): Along the vertical axis (ν_α) there are two doublets with splittings $J_{\alpha\beta}$ and $J_{\alpha\beta'}$. Along the horizontal axis (ν_β) there are three successive doublings to give 8 peaks. The splittings are $J_{\alpha\beta}, J_{\beta\beta'}, J_{\beta\gamma}$.

$H_\alpha - H_{\beta'}$ crosspeak (32 subpeaks): The cross peak shape is same as H_α - H_β but with J values corresponding to β'.

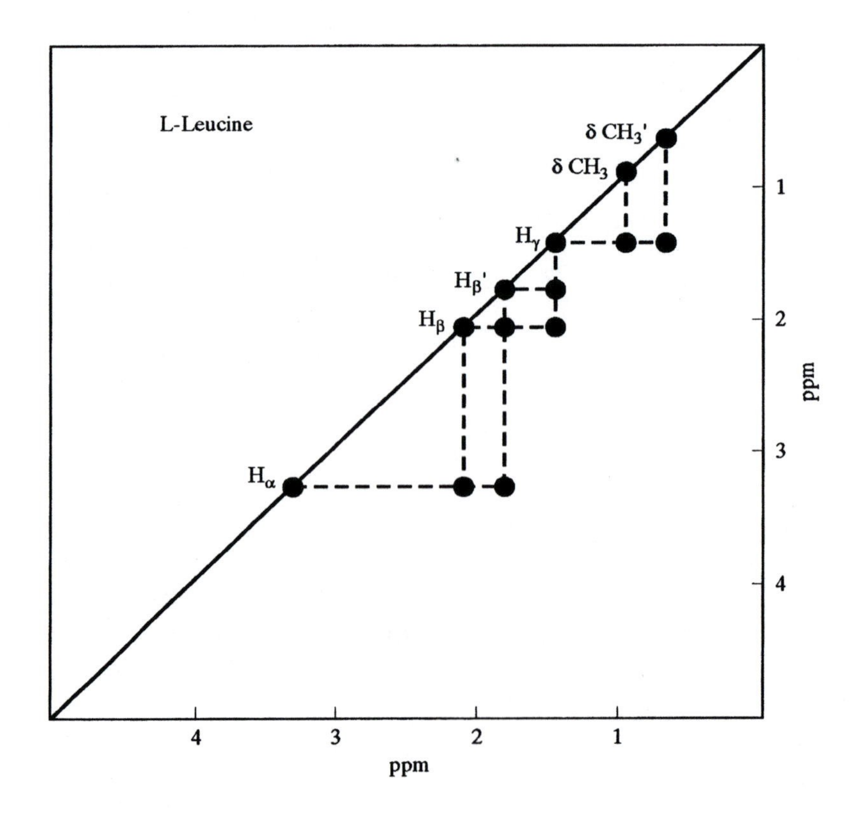

H_β - $H_{\beta'}$ crosspeak (64 subpeaks): Along the vertical axis (v_β) there are 8 peaks with splittings $J_{\alpha\beta}, J_{\beta\beta}, J_{\beta\gamma}$. Along the horizontal axis (v_β) there are 8 peaks with splittings $J_{\alpha\beta'}, J_{\beta\beta'}, J_{\beta'\gamma}$.

H_β - H_γ crosspeak (512 subpeaks): Along vertical axis (v_β) there are 8 peaks with splittings $J_{\alpha\beta}, J_{\beta\beta'}, J_{\beta\gamma}$. Along the horizontal axis (v_γ) there are $64 = 2^*2^*4^*4$ peaks with splittings $J_{\gamma\beta}, J_{\gamma\beta'}, J_{\gamma\delta}, J_{\gamma\delta'}$.

H'_β - H_γ crosspeak (512 subpeaks): The cross peak shape is same as H_β - H_γ , but with J values corresponding to β' .

H_γ - H_δ crosspeak (256 subpeaks): Along the vertical axis (v_γ) there are 64 peaks with splittings $J_{\gamma\beta}, J_{\gamma\beta}, J_{\gamma\delta}, J'_{\gamma\delta}$. Along the horizontal axis (v_δ) there is a quartet with splitting $J_{\delta\gamma}$.

H_γ - H'_δ crosspeak (256 subpeaks): The cross peak shape is same as H_γ - H_δ , but with J values corresponding to δ' .

b) Isoleucine

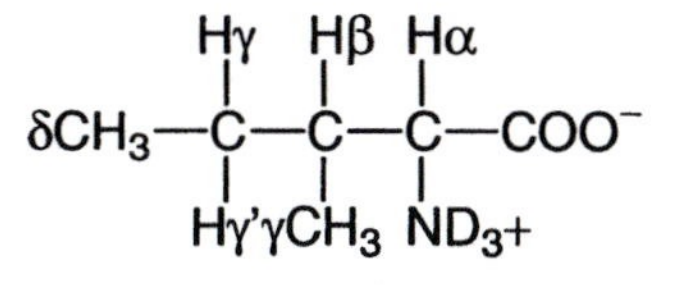

H_α - H_β crosspeak (64 subpeaks): Along the vertical axis (v_α) there is one doublet with splitting $J_{\alpha\beta}$. Along the horizontal axis (v_β) there are $32 = 2^*2^*2^*4$ peaks. The splittings are $J_{\alpha\beta}, J_{\beta\gamma}, J'_{\beta\gamma}, J_{\beta\gamma\text{CH3}}$.

H_β - H_γ crosspeak (512 subpeaks): Along the vertical axis (v_β) there 32 peaks with splittings $J_{\alpha\beta}, J_{\beta\gamma} J'_{\beta\gamma}, J_{\beta\gamma\text{CH3}}$. Along the horizontal axis (v_γ) there are $16 = 2^*2^*4$ peaks. The splittings are $J_{\gamma\beta}, J'_{\gamma\gamma}, J_{\gamma\delta\text{CH3}}$,.

H_β - H'_γ crosspeak (512 subpeaks): The cross peak shape is same as H_β - H_γ but with J values corresponding to γ .

H_β - $H_{\gamma\text{CH3}}$ crosspeak (32 subpeaks): Along the vertical axis (v_β) there are $16 = 2^*2^*4$ peaks with splittings $J_{\beta\gamma}, J'_{\beta\gamma}, J_{\beta\gamma\text{CH3}}$. Along the horizontal axis ($v_{\gamma\text{CH3}}$) there is a doublet with splitting $J_{\beta\gamma\text{CH3}}$.

H_γ - H'_γ crosspeak (256 subpeaks): Along the vertical axis (ν_γ) there are $16 = 2^*2^*4$ peaks with splittings $J_{\gamma\beta}, J'_{\gamma\gamma}, J_{\gamma\delta CH3}$. Along the horizontal axis (ν'_γ) there are 16 peaks with splitting $J_{\gamma'\beta} J'_{\gamma\gamma}, J_{\gamma'\delta CH3}$.

H_γ - $H_{\delta CH3}$ crosspeak (64 subpeaks): Along the vertical axis (ν_γ)there are $16 = 2^*2^*4$ peaks with splittings $J_{\gamma\beta}, J'_{\gamma\gamma}, J_{\gamma\delta CH3}$. Along the horizontal axis ($\nu_{\delta CH3}$) there are 2 doublets with splittings $J_{\delta CH3\gamma}$ and $J'_{\delta CH3\gamma}$.

H'_γ - $H_{\delta CH3}$ crosspeak (64 subpeaks): The cross peak shape is same as H_γ - $H_{\delta CH3}$ but with Jvalues corresponding to γ'.

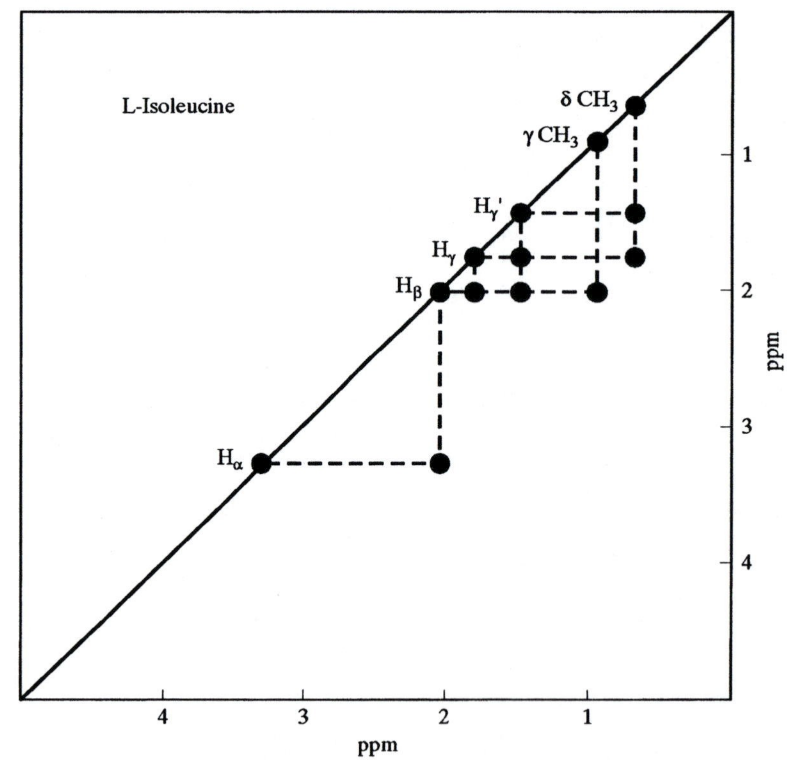

c) The patterns of cross peaks are distinctly different for leucine and isoleucine. For example there are three cross peaks at ν_β for isoleucine and only two for leucine. At the lower resolution we would see two cross peaks at ν_β for isoleucine and only one for leucine.

32. a) The DNA oligonucleotide is self complementary, so it can form a double helix at10°C. The single resonance corresponds to 100% duplex. As the temperature is raised the duplex undergoes a transition to another form, probably a hairpin. The duplex and hairpin are in slow exchange so two distinct peaks are seen. Above 35°C the hairpin melts to single strand. The kinetics are fast so only one peak is soon at the weighted average of the chemical shift of the methyl in the hairpin and single strand. By 70°C there is 100% single strand.

33. a) $$\Delta G^{\circ} = -RT\ln K = -RT\ln\frac{[\text{form I}]}{[\text{form II}]} = RT\ln\left(\frac{3}{1}\right)$$
$$= -(0.008314\ \text{kJ K}^{-1}\ \text{mol}^{-1})(298)\ln 3$$
$$= \underline{-2.72\ \text{kJ mol}^{-1}}$$

b) In the trans form the δ protons of proline are near the $\alpha - H$ of the previous amino acid; in the cis form they are far away. Measure NOEs from proline δ protons to $\alpha - H$ to identify trans δ protons. Vary temperature to learn which other peaks change with trans protons.

c) The rate of cis-trans isomerization is slow. Chemical shift difference is approximately 0.1 ppm. Therefore rate of isomerization must be slower than $(0.1)(400) = 40\ \text{s}^{-1}$.

33. d) If the two δ protons are assumed to have the same chemical shift, then we will see a triplet caused by the two δ protons (assumed equivalent). If the two δ protons have distinguishable chemical shifts, each will be a doublet of triplets. The doublet splitting is the two-bond δ - δ splitting.

CHAPTER 11

1. We assume that the two carboxyl groups are so far apart that the identical and independent sites model is applicable. The reaction

$$HO_2C-R-CO_2H \longleftrightarrow {}^-O_2C-R-CO_2H + H^+$$

can be represented by

$$11 \longleftrightarrow 10 + H^+; \quad K = [10][H^+]/[11]$$

and

$$11 \longleftrightarrow 01 + H^+; \quad K = [01][H^+]/[11]$$

where 0 and 1 represent protonated and unprotonated carboxyl groups, respectively.

The equilibrium constant K_1 is

$$K_1 = \frac{([10]+[01])[H^+]}{[11]} = 2K$$

The equilibrium constant K_2 is

$$K_2 = \frac{[00][H^+]}{[10]+[01]}$$

Because

$$\frac{[00][H^+]}{[10]} = K \text{ and } \frac{[00][H^+]}{[01]} = K, \text{therefore } K_2 = K/2$$

We now see that $k_2/K_1 = 4$.

2. We use 1's to represent occupied sites and 0's to represent unoccupied sites.

	species	statistical weight
a) Independent sites	1100	S^2
	1010	S^2
	1001	S^2
	0110	S^2
	0101	S^2
	0011	S^2
b) Nearest neighbor interactions	1100	τS^2
	1010	S^2
	1001	S^2
	0110	τS^2
	0101	S^2
	0011	τS^2
c) Exclusion of nearest neighbour	1010	S^2
	1001	S^2
	0101	S^2

Other species have zero statistical weight.

d) For case (b), the statistical weight for the species $PA_4(1111)$ is $\tau^3 S^4$. The statistical weight for the species PA_2 is, from the tabulation above, $3(S^2 + \tau S^2)$. Therefore, $[PA_4]/[PA_2] = \tau^3 S^4/3(S^2 + \tau S^2) = \tau^3 S^2/3(1+\tau)$.

3. Ring formation will predominate when the average distance between the ends of the same molecule is much smaller than the average distance between the ends of different molecules. The former is expected to be nearly concentration independent; the latter is expected to be inversely proportional to concentration. Thus, at concentrations sufficiently low, ring formation will predominate; and at concentrations sufficiently high, intermolecular aggregation will predominate. We have given a more quantitative discussion in Example 11.12.

4. According to the hydrogen bonding pattern in an alpha helix, the first four carbonyl oxygens are always represented by 0's. We tabulate below the allowed states of the four remaining residues. (States such as 1011 or 1001 are not allowed).

Configuration	Statistical weight
1000	σs
0100	σs
0010	σs
0001	σs
1100	σs^2
0110	σs^2
0011	σs^2
1110	σs^3
0111	σs^3
1111	σs^4

$$Z = 4\sigma s + 3\sigma s^2 + 2\sigma s^3 + \sigma s^4$$

$$= \sigma s(4 + 3s + 2s^2 + s^3)$$

$$\nu = \frac{4\sigma s + 2\cdot 3\sigma s^2 + 3\cdot 2\sigma s^3 + 4\sigma s^4}{Z}$$

$$= \frac{4 + 6s + 6s^2 + 4s^3}{4 + 3s + 2s^2 + s^3}$$

5. Following the rules, two protons bound to adjacent carboxyl groups will have statistical weight τs^2.

a)

Species	Statistical weight	Species	Statistical weight
00 00	1	01 01	τS^2
10 00	S	01 10	S^2
01 00	S	00 11	τS^2
00 01	S	11 01	$\tau^2 S^3$
00 10	S	01 11	$\tau^2 S^3$
11 00	τS^2	10 11	$\tau^2 S^3$
10 01	S^2	11 10	$\tau^2 S^3$
10 10	τS^2	11 11	$\tau^2 S^3$

b) $$Z = 1 + 4S + 2S^2 + 4\tau S^2 + 4\tau^2 S^3 + \tau^4 S^4$$

$$\nu = \frac{S}{Z}\left(\frac{\delta Z}{\delta S}\right)_\tau = \frac{4S + 4S^2 + 8\tau S^2 + 12\tau^2 S^3 + 4\tau^4 S^4}{Z}$$

$$= \frac{4S(1 + S + 2\tau S + 3\tau^2 S^2 + \tau^4 S^4)}{1 + 4S + 2S^2 + 4\tau S^2 + 4\tau^2 S^3 + \tau^4 S^4}$$

c)

pH	S	Z	f	f (single COO^-)
4	10	254.4	0.46	0.91
5	1	7.04	0.29	0.5
6	0.1	1.42	0.079	0.091

6. The first four carbonyl oxygens are always represented by 0's due to the hydrogen bonding pattern of an alpha-helix. The configuration of the polypeptide can be described by a sequence of 46 digits, where sequences like …1011… or …10011… are not allowed. The sequence of 46 consecutive 1's is given statistical weight unity.

a) Statistical weight for a belix $= \sigma s^{46}$

b) For the species with sequences containing 4, 6 and 10 amides in helical segments, separated by at least 3 non-helical amino acids

statistical weight $= \sigma^3 s^{20}$

c) Because $\sigma = 10^{-4}$, any term containing factors of $\sigma, \sigma^2, \sigma^3$, etc. will be exceedingly small relative to the sequence of all zeros (coil configuration) which has statistical weight $= 1$.

d) Any term containing σ^2, σ^3 etc. will be small compared with those containing σ as a factor times the same (or smaller) power of s. Thus, the largest statistical weight will be for the α - helix, corresponding to a string of 46 consecutive 1's, which will be $\sigma s^{46} = (10^{-4})(10^{46}) = 10^{42}$.

e) Because ΔH^o is negative for helix $\longleftrightarrow$ coil,

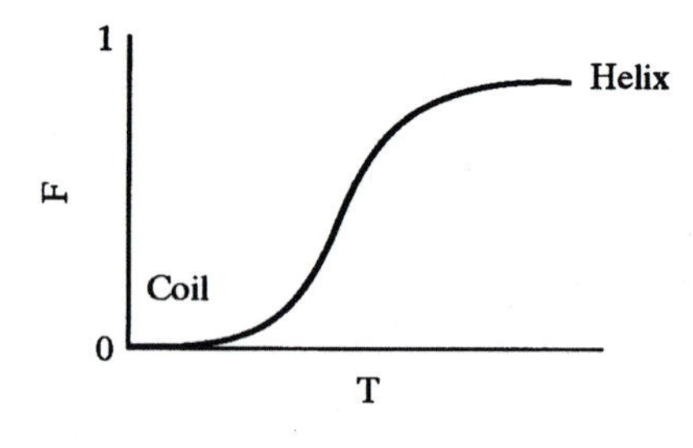

f) $f = \dfrac{\nu}{49} = \dfrac{1}{49} \cdot \dfrac{\sigma s^{46}}{Z}$

7. a) We use 0's and 1's to represent unpaired and paired AU's, counting from the end without the loop.

No. of base pairs	Species	Statistical weight
6	111111	$\sigma_7 s^6$
5	011111	$\sigma_7 s^5$
	111110	$\sigma_9 s^5$
4	001111	$\sigma_7 s^4$
	011110	$\sigma_9 s^4$
	111100	$\sigma_{11} s^4$
3	000111	$\sigma_7 s^3$
	001110	$\sigma_9 s^3$
	011100	$\sigma_{11} s^3$
	111000	$\sigma_{13} s^2$
2	000011	$\sigma_7 s^2$
	000110	$\sigma_9 s^2$
	001100	$\sigma_{11} s^2$
	011000	$\sigma_{13} s^2$
	110000	$\sigma_{15} s^2$
1	000001	$\sigma_7 s$
	000010	$\sigma_9 s$
	000100	$\sigma_{11} s$
	001000	$\sigma_{13} s$
	010000	$\sigma_{15} s$
	100000	$\sigma_{17} s$
0	000000	1

b) The mole fraction of each species is equal to its statistical weight divided by the function

$$Z = \sigma_7 s^6 + s^5(\sigma_7 + \sigma_9) + s^4(\sigma_7 + \sigma_9 + \sigma_{11}) + s^3(\sigma_7 + \sigma_9 + \sigma_{11} + \sigma_{13}) + s^2(\sigma_7 + \sigma_9 + \sigma_{11} + \sigma_{13} + \sigma_{15}) + s(\sigma_7 + \sigma_9 + \sigma_{11} + \sigma_{13} + \sigma_{15} + \sigma_{17}) + 1$$

c) The absorbance A is

$$A = l\varepsilon_0 c(1/Z) + l\varepsilon_1 c\left(\sum_{3}^{8} \sigma_{2j+1}\, s/Z\right) + l\varepsilon_2 c\left(\sum_{3}^{7} \sigma_{2j+1}\, s^2/Z\right) + \cdots$$

$$= \frac{cl}{Z}\left[\varepsilon_0 + s\varepsilon_1 \sum_{3}^{8} \sigma_{2j+1} + s^2\varepsilon_2 \sum_{3}^{7} \sigma_{2j+1} + s^3\varepsilon_3 \sum_{3}^{6} \sigma_{2j+1} + \cdots\right.$$

$$\left. + s^6\varepsilon_6 \sum_{3}^{3} \sigma_{2j+1}\right]$$

where l is the path length and Z is the sum defined in (b).

8. Contour length $L = 0.34\times10^{-9}\,\text{m}\times10^{7}$

$$= 0.34\times10^{-2}\,\text{m}$$
$$= 3.4\ \text{mm}$$

Number of segments $N = L/100\times10^{-9}\,\text{m}$

$$= 0.34\times10^{-2}/10^{-7}$$

$$= 3.4\times10^{4}$$

Mean square end-to-end distance $\langle h^2 \rangle = N(100\ \text{nm})^2$

$$= 3.4\times10^{4}(10^{-7}\,\text{m})^2$$
$$= 3.4\times10^{-10}\ \text{m}^2$$
$$= 3.4\times10^{8}\,\text{nm}^2$$

9. The allowed energy levels are

$$E = (n_x^{\ 2} + n_y^{\ 2} + n_z^{\ 2})\frac{h2}{8\ \text{ms}^2}$$

Energy level	(n_x, n_y, n_z)	Degeneracy
1	1,1,1	1
2	2,1,1	3
	1,2,1	
	1,1,2	
3	2,2,1	3
	2,1,2	
	1,2,2	
4	3,1,1	3
	1,3,1	
	1,1,3	
5	2,2,2	1

10. a) There are 26 ways of selecting each of the 3 letters. The total number of ways is, therefore, $26^3 = 17,576$. Thus, it is possible to make up 17,576 three-lettered words from AAA to ZZZ.

b) The number of different ways of choosing 5 players from a group of 100 people is $100!/(5!95!) = 7.5\times10^7$

c) The number of different proteins is $20^{100} = 1\times10^{130}$.

11. Although He atoms obey Bose-Einstein statistics and electrons Fermi-Dirac statistics, for our particle-in-the-box problem the system is very dilute—only one particle and many energy levels— so we can apply Boltzmann statistics.

$$\frac{N_2}{N_1} = \frac{g_2}{g_1}\exp[-E_2 - E_1)/kT \qquad \text{(Eq.11.68)}$$

$$E = (n_x^{\ 2} + n_y^{\ 2} + n_z^{\ 2})\frac{h^2}{8\,\mathrm{ma}^2}$$

Following the method of solving Problem 11.9 for E_1,
$n_x^{\ 2} + n_y^{\ 2} + n_x^{\ 2} = 3$ and $g_1 = 1$; for E_2, $n_x^{\ 2} + n_y^{\ 2} + n_z^{\ 2} = 6$ and $g_2 = 3$.

$$E_2 - E_1 = 3\,\mathrm{h}^2/8\,\mathrm{ma}^2$$

$$\frac{N_2}{N_1} = 3\exp[-3\,\mathrm{h}^2/8\,\mathrm{ma}^2\mathrm{kT}] = 3\exp[-1.192\times10^{-44}/\mathrm{ma}^2T]$$

For He

$$m = 4\times10^{-3}/6.023\times10^{23};\, a = 10^{-2}\,\mathrm{m}$$

$$\frac{\mathrm{N}_2}{\mathrm{N}_1} = 3\exp[-1.795\times10^{-14}/T]$$

At 10 K, 1000 K or 10,000 K, the exponential term is essentially 1 and

$$N_2/N_1 = 3$$

For the electron case

$$m = 9.109\times10^{-31}\,\mathrm{kg}\; a = 10^{-9}\,\mathrm{m} = 3\exp[-13100/T]$$

$$\frac{N_2}{N_1} = 3\exp[-13100/T]$$

a) $T = 10\,\mathrm{K}$

$$N_2/N_1 = 3\exp[-1310.9] = 0$$

b) $T = 1000\text{ K}$

$N_2/N_1 = 3\exp[-13.1] = 6.14\times 10^{-6}$

c) $T = 10{,}000\text{ K}$

$N_2/N_1 = 3\exp[-1.31] = 0.810$

12. a) We can think of the problem as blindly picking out 100 pennies from a well mixed bagful. The number of ways that all 100 pennies picked are heads up is $t_1 = 1$; that is, when each and every penny picked is a head. Similarly, the number of ways that all 100 pennies picked are tails is $t_2 = 1$. The entropy change when 100 pennies are changed from all heads to all tails is

$$\Delta S = k\ln(t_2/t_1) = 0$$

b) The number of ways t_3 of picking 100 pennies with 50 heads and 50 tails is

$$t_3 = \frac{100!}{50!(100-50)!} = \frac{100!}{(50!)^2}$$

Note that this is the same problem as the number of ways of taking 100 steps with 50 steps forward (Eq. 11.27). The entropy change is

$$\Delta S = k\ln(t_3/t_1)$$
$$= k\ln[100!/(50!)^2] = k[\ln 100! - 2\ln 50!]$$

We use Stirling's approximation

$\ln N! = N\ln N - N$

to calculate ln 100! and ln 50!

$\ln 100! = 360.52$

$\ln 50! = 145.60$

$$\Delta S = 69.32\text{ k} = \frac{69.32\cdot 8.314\text{ J K}^{-1}\text{ mol}^{-1}}{6.023\times 10^{23}\text{ mol}^{-1}}$$
$$= 9.57\times 10^{-22}\text{ J K}^{-1}$$

c) Before mixing, we have two separate systems, each containing one mole of pennies. If there are t_1 ways for arranging the first system and t_2 ways for arranging the second system, the total number of ways for the combined system is $t = t_1 t_2$. Because $t_1 = 1$ and $t_2 = 1, t = 1$. After mixing, the number of ways of arranging the system is

$t' = \dfrac{(2N_o)!}{N_o!N_o!}$, where N_o is Avogadro's number

$$\Delta S = k\ln(t'/t) = k[\ln(2N_o)! - 2\ln N_o!]$$

Using Stirling's approximation we have

$$\Delta S = k[2N_o \ln 2N_o - 2N_o - 2N_o \ln N_o + 2N_o]$$
$$= 2R\ln 2 = 2\cdot 8.314 \ln 2 \text{ J K}^{-1} = 11.53 \text{ J K}^{-1}$$

Note that this is the same problem as mixing 1 mole of gas A with 1 mole of gas B.

13. a) Using (Eq. 11.66) and applying it to this case

$$Z = g_1 \exp[-\varepsilon_1/kT] + g_2 \exp[-\varepsilon_2/kT]$$
$$= 2\exp[D/kT] + 1$$

b) $$< E > = \frac{1}{Z}(g_1 E_1 \exp[-E_1/kT] + g_2 E_2 \exp[-E_2/kT])$$
$$= \frac{-2D\exp[D/kT]}{2\exp[D/kT]+1} = \frac{-2D}{2+\exp[-D/kT]}$$

c) In the high temperature limit, $\exp[-D/kT] \rightarrow 1$

$$< E > = -\frac{2}{3}D$$

d) In the low temperature limit,

$$< E > = -D$$

e) $$t_i = g_i \exp[-E_i/kT]$$
$$t_2/t_1 = (g_2/g_1)\exp[-(E_2 - E_1)/kT]$$
$$= (1/2)\exp[D/kT]$$

f) In the high temperature limit, $t_2/t_1 = 1/2$

$$S = k\ln Z = k\ln(2+1) = k\ln 3$$
$$= 1.516\times 10^{-23} \text{ J K}^{-1}$$

CHAPTER 12

1. a) $$\text{Energy} = 40\times10^{3}\,\text{eV}$$
$$= 40\times10^{3}\,\text{eV}\ 1.602\times10^{-19}\ \text{J eV}^{-1}$$
$$= \underline{6.408\times10^{-15}\,\text{J}}$$

b) $$\nu = \Delta c/\text{h} = 6.408\times10^{-15}\ \text{J}/6.626\times10^{-34}\ \text{J s}$$
$$= 9.671\times10^{18}\ \text{s}^{-1}$$
$$\lambda = \text{c}/\nu = 2.998\times10^{8}\,\text{m s}^{-1}/9.671\times10^{18}\ \text{s}^{-1}$$
$$= 3.100\times10^{-11}\ \text{m} = \underline{0.31\,\text{Å}}$$

c) $$\lambda_s = c/\nu = ch/\Delta\varepsilon$$
$$\frac{(2.998\times10^{8}\,\text{m s}^{-1})(6.626\times10^{-34}\ \text{J s})}{\text{V}(1.602\times10^{-19}\ \text{J eV}^{-1}}$$
$$= \frac{1.2400\times10^{-6}}{V}(\text{in m})$$
$$= \frac{12.400}{V}(\text{in Å})$$

2. a) Energy required $= h\nu = hc/\lambda$ where λ is the wavelength corresponding to the K absorption edge.

For Cu, $\lambda = 1.380\ \text{Å} = 1.380\times10^{10}\,\text{m}$

$$\text{Energy required} = \frac{(6.626\times10^{-34}\ \text{J s})(2.998\times10^{8}\,\text{m s}^{-1})}{1.380\times10^{-10}\ \text{m}}$$
$$= \underline{1.4395\times10^{-15}\ \text{J}}$$

Similarly, for Ag, $\lambda = 0.4858\times10^{-10}$ m, and the energy required is $\underline{4.089\times10^{-15}\ \text{J}}$.

b) In general, the energy ε corresponding to a photon of wavelength λ is

$$\varepsilon = hc/\lambda = (6.626\times10^{-34}\,\text{J s})(2.998\times10^{8}\ \text{m s}^{-1})/\lambda$$
$$= 1.9865\times10^{-25}\,\text{J m}/\lambda$$
$$= 1.9865\times10^{-15}\,\text{J Å}/\lambda$$

If we express ε in units of eV,

$$\varepsilon = (1.9865\times10^{-15}\,\text{J Å})(1\,\text{eV}/1.602\times10^{-19}\ \text{J})/\lambda$$
$$= 12{,}400\,\text{eV Å}/\lambda$$

This is, of course, the same equation we obtained in Problem 12.1 (c).

The energies corresponding to
$\lambda = 1.54433$ Å (α_2), 1.54051 Å (α_2), 1.39217 Å (β) and 1.38102 Å (γ) are therefore 8029.4, 8049.3, 8906.9 and 8978.8 eV, respectively. We have already calculated the *K* shell energy level of silver to be 25, 525 eV in part (b). Thus, the energy level corresponding to the α_2 line is $25{,}525 + 8029 = 33{,}554\,\text{eV}$. Similarly, the energy levels corresponding to the other *L* shell lines are 33,574, 34,432 and 34,504 eV.

3 a) With an incident angle α_o of 90°

$$\alpha = \cos^{-1}(h/a)$$
$$\alpha = \cos^{-1}(h1.54/4)$$

For $h = 1$, $\alpha = \cos^{-1}(1.54/4) = 67.4^\circ$

For $h = 2, \alpha = \cos^{-1}(2\times1.54/4) = 39.6^\circ$

b) The heights of the first and second layer lines, measured from the apex of the cones, are

$$y_1 = \frac{3\,\text{cm}}{\tan 67.4^\circ} = 1.249\,\text{cm}$$
$$y_2 = \frac{3\,\text{cm}}{\tan 39.6^\circ} = 3.626\,\text{cm}$$

The distance between these layer lines is $y_2 - y_1 = \underline{2.377\,\text{cm}}$

4. In the direction of the source beam, the distance between the two points 1 and 2 is $R\cos\phi$, where ϕ is defined as in the problem.

In the direction of the detector, the difference in the path is $R\cos(\theta-\phi)$. Therefore,

$$\Delta = R\cos\phi - R\cos(\theta-\phi)$$

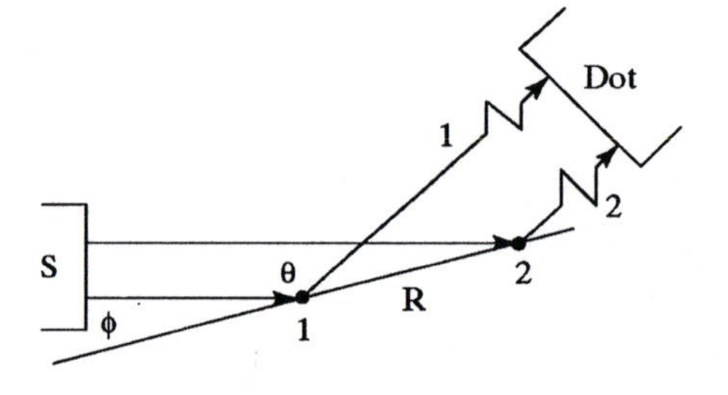

5. Using Eq. (12.5) to characterize the scattering from two points we require that the term $\frac{2\pi\Delta}{\lambda} > \cos^{-1}0.01$, where $\lambda = 1.54$ Å $\frac{2\pi\Delta}{\lambda} > \cos^{-1}0.01 = 89.4^\circ = 1.561$ radians $\Delta > 1.54$ Å $1.56/2\pi = 0.383$ Å

6. For $n = 1$, the Bragg angle is, from Eq. (12.9)
$\theta = \sin^{-1}(\lambda/2d_{hkl}) = \sin^{-1}(1.54\,\text{Å}/2d_{hkl})$

If $d_{hkl} = 5$ Å, $\theta = 8.86^\circ$
If $d_{hkl} = 10$ Å, $\theta = 4.42^\circ$
If $d_{hkl} = 1000$ Å, $\theta = 0.044^\circ$

7. The spacing corresponding to $\theta = 90^\circ$ or $\sin\theta = 1$ is

$$d_{hkl} = \frac{\lambda}{2} = 1.54\,\text{Å}/2 = 0.77\,\text{Å}$$

The theoretical limit of resolution is thus 0.77 Å with 1.54 Å wavelength X-rays.

8. a) For 1 mole of unit cells, the total volume V is $N_o(5.64\,\text{Å})^3$ and the total mass is $N_o(2.163\text{ g cm}^{-3})(5.64\,\text{Å})^3(10^{-8}\text{ cm/Å})^3$. Because 1 mole of unit cells contains 4 moles of Na^+ and 4 moles of Cl^-, the total mass must be 4(58.45 g). Thus, $N_o(2.163\text{ g cm}^{-3})(5.64\times10^{-8}\text{ cm})^3 = 4(58.45\text{ g})$, and $N_o = \underline{6.025\times10^{23}}$.

b) The total mass of lysozyme in 1 mole of unit cells is

$$m = (6.022\times10^{23})(79.1)^2(37.9)(10^{-24}\text{ cm}^3)(1.242\text{ g cm}^{-3})(0.644)$$
$$= 1.142\times10^5\text{ g}$$

There are 8 mol of lysozyme molecules per mol of unit cells. The mol wt of lysozyme is, therefore, $1.142\times10^5/8=\underline{1.428\times10^4}$.

9 a) Consider the group of lattice lines in the lower left of Fig. 12.11. We notice that the interplanar distance is equal to the distance from the origin at the lower left corner to the line next to it. We recall that the intercepts of the line of Miller indices (h, k) next to the origin is a/h and b/k. For a three dimensional lattice, the intercepts of a corresponding plane are $a/h, b/k$ and c/l.

Because $\overline{\mathrm{OP}}=d_{hkl}$ is perpendicular to the plane,

$$\cos\alpha=\frac{d_{hkl}}{(a/h)},\cos\beta=\frac{d_{hkl}}{(b/k)},\cos\gamma=\frac{d_{hkl}}{(c/l)}$$

Adding the squares of the left and right sides of the equations, we obtain

$$\cos^2\alpha+\cos^2\beta+\cos^2\gamma=d_{hkl}^2\left[\left(\frac{h}{a}\right)^2+\left(\frac{k}{b}\right)^2+\left(\frac{l}{c}\right)^2\right]$$

But $\cos\alpha, \cos\beta$ and $\cos\gamma$ are the direction cosines of the line $\overline{\mathrm{OP}}$ and, therefore, the sum of their squares is equal to 1. Thus,

$$d_{hkl}^2=\frac{1}{[(h/a)^2+(k/b)^2+(l/c)^2]}, \text{ or}$$

$$d_{hkl}=\frac{1}{[(h/a)^2+(k/b)^2+(l/c)^2]^{1/2}}$$

b) For a cubic lattice, $a=b=c$ and

$$d_{hkl}=\frac{a}{[h^2+k^2+l^2]^{1/2}}$$

Substituting the values of (h, k, l) into the above equation gives immediately the ratios 1:0.707:0.578.

10. a) Because the atom is at the origin

$X_1 = 0, Y_1 = 0, Z_1 = 0$

Let f_1 be the atomic scattering factor of the atom.

From Eqs. (12.24 a) and (12.24 b),

$A(h,k,l) = f_1 \cos[2\pi(0)] = f_1$

for any values of (h, k, l).

$B(h,k,l) = f_1 \sin[2\pi(0)] = 0$
for any values of (h, k, l)
$A^2 + B^2 = f^2$, independent of the values of (h, k, l). Therefore, the intensities of the diffractions by the planes are all equal.

b) Let f_1 be the atomic scattering factor of the atom.

$X_1 = 0, Y_1 = 0, Z_1 = 0$ and $X_2 = \frac{1}{2}, Y_2 = \frac{1}{2}, Z_2 = \frac{1}{2}$
From Eqs. (12.24 a) and (12.24 b)

$$A = f_1 \cos(0) + f_1 \cos\left[2\pi\left(\frac{h}{2} + \frac{k}{2} + \frac{l}{2}\right)\right]$$
$$= f_1 + f_1 \cos[\pi(h + k + l)]$$
$$B = f_1 \sin(0) + f_1 \sin[\pi(\mathrm{h} + \mathrm{k} + \mathrm{l})]$$
$$= 0 + 0 = 0$$

(The sum $h + k + l$ is an integer. The sine of an integral number times π is always 0).
Therefore, the intensity is given by A^2,
or $f_1^{\,2}[1 + \cos\{\pi(h + k + l)\}]^2$
For $(h, k, l) = (1, 0, 0), A^2 = f_1[1 + \cos(\pi)]^2 = 0$
For $(h, k, l) = (2, 0, 0), A^2 = f_1^{\,2}[1 + \cos(2\pi)]^2 = 4f_1^{\,2}$
For $(h, k, l) = (1, 1, 0), A^2 = f_1^{\,2}[1 + \cos(2\pi)]^2 = 4f_1^{\,2}$
The intensity of the (1, 0, 0) reflection is zero, and the intensities of the (2, 0, 0) and (1, 1, 0) reflections are equal.

11. a)

b)

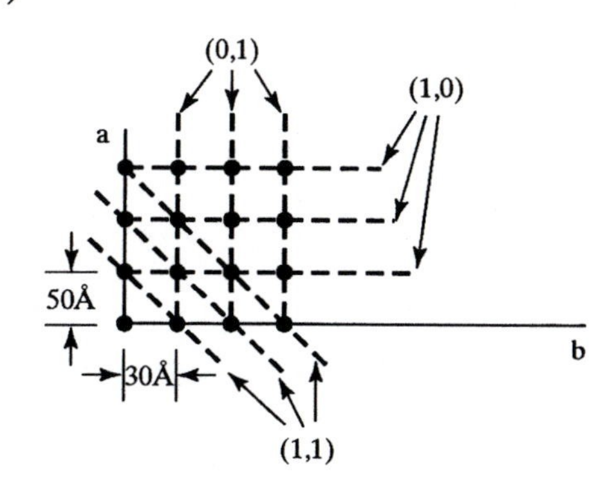

c)

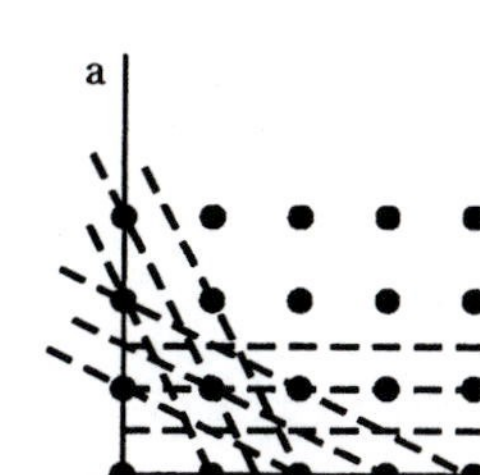

11. d, e) $d_{h,k} = \left(\frac{h^2}{a^2} + \frac{k^2}{b^2}\right)^{-1/2}$ $a = 50\text{ Å},\ \ B = 30\text{ Å}\ \sin\theta = \lambda/2d_{hk}$

h	k	d_{hk}(Ǎ)	sin θ	θ(degree)
1	0	50.00	0.0154	0.882
0	1	30.00	0.0257	1.47
1	1	25.72	0.0299	1.71
2	0	25.00	0.0308	1.76
1	2	14.37	0.0535	3.07
2	1	19.21	0.0400	2.30
2	2	12.86	0.0599	3.43

f) $A(h,k) = \Sigma f_j \cos[2\pi(h X_j + k Y_j)]$

$B(h,k) = \Sigma f_j \sin[2\pi(hX_j + kY_j)]$

$$\alpha(h,k) = \tan^{-1}\frac{B(h,k)}{A(h,k)}$$

$$X_j = X_j/50 Y_j = Y_j/30$$

$$A(1,0) = 82[\cos 2\pi(12/50) + \cos 2\pi(12/50)]$$
$$= 82[0.0628 - 0.0628]$$
$$A(1,0) = 0$$
$$B(1,0) = 82[\sin 2\pi(12/50) + \sin 2\pi(37/50)]$$
$$= 82[0.9980 - 0.9980]$$
$$B(1,0) = 0$$
$$I(1,0) = 0$$
$$A(2,0) = 82[\cos 2\pi(24/50) + \cos 2\pi(74/50)] = 82[-0.9921 - 0.9921]$$
$$A(2,0) = -162.7$$
$$B(2,0) = 82[\sin 2\pi(24/50) + \sin 2\pi(74/50)] = 82[0.1253 + 0.1253]$$
$$B(2,0) = 20.5$$
$$\alpha(2,0) = \tan^{-1}(20.5/-162.7)$$
$$\alpha(2,0) = -0.125 \text{ radians} = -7.16 \text{ degrees}$$
$$I(2,0) = A^2 + B^2 = 2.69\times10^4$$

The Pb atoms are very close to positions 1/4 and 3/4 of the unit cell along the a axis. This means that their scattering amplitude are out of phase with the periodicity of the unit cell (1, 0 planes), but in phase with half that periodicity (2, 0 planes). Their separation of 25 Å is half the separation of 1, 0 planes and equal to the separation of the 2, 0 planes.

12. a) Using Eqs. (12.34) and (12.35) for $V = 10^4$ V

$$\frac{\text{Ve}}{2m_e c^2} = \frac{(10^4 \text{ eV})(1.602\times10^{-19} \text{ J eV}^{-1})}{2(9.107\times10^{-31} \text{ kg})(2.998\times10^8 \text{ m s}^{-1})^2} = 9.79\times10^{-3}$$

$$V_{\text{corrected}} = 10^4(1 + 9.79\times10^{-3}) \text{ V} = 10098 \text{ V}$$

$$\lambda = h/\sqrt{2m_e \text{ eV}_{\text{corrected}}} = \frac{6.626\times10^{-34} \text{ J s}}{\sqrt{(2)(9.107\times10^{-31} \text{ kg})(10098 \text{ eV})(1.602\times10^{-19} \text{ J eV}^{-1})}}$$

$$= 1.220\times10^{-11} \text{ m} = 0.122 \text{ Å}$$

b) Similarly, for $V = 10^5$ volts

$$V_{\text{corrected}} = 109{,}790 \text{ V and}$$

$$\lambda = \underline{3.702\times10^{-12} \text{ m} = 0.03702 \text{ Å}}$$

13. Let m be the mass of a neutron. The kinetic energy of 1 mol of thermal neutrons at 300 K is $3\,RT/2$. The average kinetic energy of 1 neutron at this temperature is

$$\frac{p^2}{2m} = \frac{3RT}{2N_o} = \frac{3kT}{2}$$

$$\lambda = h/p = h/\sqrt{3\text{mkT}}$$

$$= \frac{6.626\times10^{-34} \text{ Js}}{\sqrt{3(1.673\times10^{-27} \text{ kg})(1.3805\times10^{-23} \text{ J K}^{-1})(300K)}}$$

$$= \underline{1.45\times10^{-10} \text{ m} = 1.45 \text{ Å}}$$

14. a) to c) See appropriate sections in text

d) Most of the protein and nucleic acid crystals contain a high proportion of water so the molecular structure may be very similar to the biologically relevant conditions.

15. a)

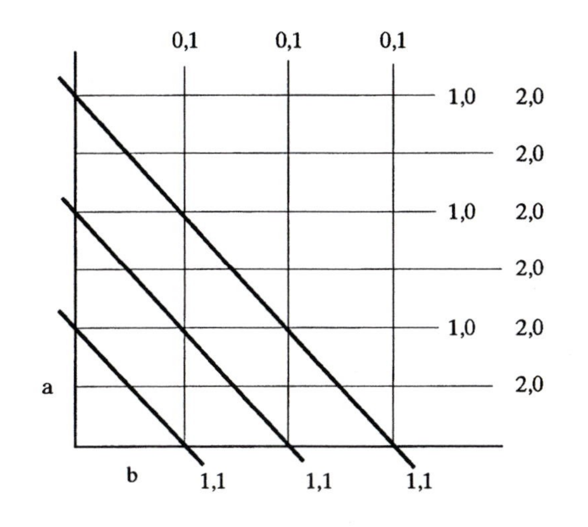

b) $d_{hk} = [(h^2/a^2) + k^2/b^2]^{-\frac{1}{2}}$

$d_{0,1} = 90$ Å

$d_{1,0} = 80$ Å

$d_{2,0} = 40$ Å

$d_{1,1} = 36.55$ Å

c) $\sin\theta = \lambda/2d = 1.75/(2)(7.00) = 0.1250$

$= 0.1253$ radians

d) $A(2,0) = 74\cos 2\pi(20/80) = 74\ \cos(\pi/2) = 0$

$B(2,0) = 74\sin(\pi/2) = 74$

$\alpha(2,0) = \tan^{-1}(B/A) = \tan^{-1}(\infty) = 90^\circ$

$I(2,0) = A^2 + B^2 = 5476$

16. a) path difference $= n\ \lambda = 0, 1.54$ Å, 3.08 Å, 4.62 Å

b) path difference $= (2n-1)/(\lambda/2) = 0.77$ Å, 2.31 Å, 3.85 Å

c) $d = \lambda/2\sin\theta = 8.83$ Å

d) $d_{hkl} = [(h/a)^2 + (k/b)^2 + (l/c)^2]^{-1/2}$

Maximum $d = d_{100} = d_{010} = d_{001} = 5$ Å

e) $d_{\text{minimum}} = \lambda/2 = 0.77$ Å

f) $F_h = f_c \Sigma e^{2\pi i h X j}\ h = \text{integer}$

$F_1 = 6\ [e^{2\pi i(0)} + e^{2\pi i(2/3)}]$

$= 6\ [1 + \cos(4\pi/3) + i\sin(4\pi/3)]$

$F_1 = 6\ [1 - 0.50 - i\,0.866]$

$I_1 = F_1^2 = 6^2[0.5^2 + 0.866^2] = 36$

$\alpha_1 = \tan^{-1}(0.5/-0.866) = -0.7137$ radians

17. a) A unit cell for a crystal is a fundamental volume which by translation along the three crystal axes can reproduce the entire crystal. All cells shown are correct unit cells.

b) No. The crystal complex is not necessarily the same as the complex in solution.

c) The diffraction pattern will have the symmetry of the unit cell.

d) The spacing would increase because the interplanar distances would decrease.

e) The diffraction pattern spacing would double.

f) The diffraction pattern spacing would not change.